静物创意压花艺术

[美] 朱少珊 著 \ 世界压花协会推荐参考书

图书在版编目（CIP）数据

静物创意压花艺术 /（美）朱少姗著. -- 北京：
中国林业出版社，2021.5

ISBN 978-7-5219-1161-9

Ⅰ.①静… Ⅱ.①朱… Ⅲ.①压花—装饰美术—
技法(美术) Ⅳ.①J525.1

中国版本图书馆CIP数据核字(2021)第095456号

版权备案号：01-2021-2589号

策划编辑：印芳

责任编辑：印芳　邹爱

出　　版	中国林业出版社（100009 北京西城区刘海胡同7号）
电　　话	010-83143565
发　　行	中国林业出版社
印　　刷	河北京平诚乾印刷有限公司
版　　次	2021年6月第1版
印　　次	2021年6月第1次印刷
开　　本	710毫米 × 1000毫米　1/16
印　　张	10
字　　数	164千字
定　　价	69.00元

前言

Preface

  自从我的初级、中级和高级压花艺术三本书出版了之后，一直试图写关于创意压花艺术和关于从高级迈向大师级压花艺术方法和技巧的书。对我而言，这是一项非常艰巨的任务。我希望这套书《风景创意压花艺术》《静物创意压花艺术》能往压花的艺术方面引导我的读者，而不仅仅是简单的作品制作方法。

  压花艺术有什么用？艺术让我思考，让我去探索和想象。使我欣赏到美学价值。约翰·拉伯克（John Lubbock）写道："毫无疑问，艺术是人类幸福中最纯粹，最高的元素之一。它通过眼睛训练大脑，并且通过大脑训练眼睛。如同太阳赋予花色彩，艺术赋予生活色彩。"我希望压花艺术能为每个人的生活增色。

  在我写作期间，我先生庆承侃为我写了首诗。在此分享一下。他的鼓励，是我坚持写出这套书的一大动力。并且感谢他为我的压花作品提出自己的看法和意见。

> 纤手玉指理奇葩，
> 花草枝叶尽入画。
> 当年黛玉若识君，
> 红楼梦里无葬花。

  书里通过不同的范例课程和图片，引导读者试验压花艺术不同的创意和各种方法还有技巧并掌握它们。因此，每个人都可以使用这些方法和技巧来创作出属于自己的作品。

After publishing three beginner to advanced pressed flower art books, I had been trying to write a book on an even higher level of methods and techniques. It has been a very difficult task for me. I want this book to guide my readers on the art side of pressed flowers. This is not a simple how-to book.

What is pressed flower art for? Art makes me think, enables me to explore, and allows me to imagine. It makes me appreciate the value of aethetics. John Lubbock wrote, "Art is unquestionably one of the purest and highest elements in human happiness. It trains the mind through the eye, and the eye through the mind. As the sun colors flowers, so does art color life.". I hope that pressed flower art colors everyone's lives.

This book contains many step-by-step projects and sample pictures. My goal is to guide readers to experiment with all kinds of creative ideas, to try out the methods and techniques, and to master them. Thus, everyone can create their own artwork with these methods and techniques.

I want to thank my husband for the encouragement and support during my writing.

2021.04

前言
Preface 003

压花艺术创意和技巧
Pressed Flower Art Ideas and Techniques 008
压花艺术欣赏 Pressed Flower Art Appreciation 011

形状
Form 016

几何形状 Geometric Shape 018
植物画 Cropped botanical 019
运用圆形制作一个完美的花环
Using Round Shape to Make a Perfect Wreath 023
　材料 Materials 023
　制作圆形状 Making the Round Shape Guide 024
　制作花环 Making the Wreath 025
　变化步骤 Variation of the Steps 026
　变化花材 Variation of the materials 026
　变化框架 Variation of the Guide 026

球状 Spheres 028
　材料 Materials 029
　制作叶脉 Making Skeleton Leaves 029
　制作球形花瓶 Making the Ball Shaped Vases 029

长方盒状 Box Shape 032
　材料 Materials 032
　制作窗边花坛 Making the Window Box 033

混合形状 Combination of Shapes 035

组合形态 Combined Form 036
　组合大丽花 Assemble Dahlia 037

有机形状 Organic Shapes 043
　轮廓方法 Outline Method 043
　拼图方法 Puzzle Methods 044

有机形态之真实感的玫瑰 Organic Form – Realistic Roses ... 047
　玫瑰组合 Assemble Roses 048
　有角度的玫瑰组合 Assemble Roses Angle View 050
　正面的玫瑰组合 Assemble Roses Front View 054
　半开和全开的玫瑰组合 Assemble Half and Full Open Roses ... 058
　玫瑰花束制作 Making Rose Bouquets 062

目录 Contents

玫瑰花园制作 Making Rose Garden Picture　　　　　　　**070**

图案
Patterns　　　　　　　074
压花作为设计元素 Pressed Flowers as Design Elements　　076
对称 Symmetry　　　　　　　077
旋转图案 Spiral Pattern　　　　　　　078
流水图案 Flow Pattern　　　　　　　079
波浪，沙丘图案 Waves, Dune Pattern　　　　　　　080
树木，分形图案 Trees, Fractal Pattern　　　　　　　082
镶嵌图案 Tessellation Pattern　　　　　　　084
花窗玻璃式设计 Design Stained Glass Style　　　　　　　086
运用纸胶带制作简单彩色花窗玻璃设计
Easy Stained Glass with Washi Tapes　　　　　　　087
　材料 Materials　　　　　　　087
　制作方法 Procedure　　　　　　　088
彩色花窗玻璃玫瑰 Stained Glass Rose　　　　　　　091
　材料 Materials　　　　　　　091
　制作方法 Procedure　　　　　　　092
彩色花窗玻璃模板 Stained Glass Pattern　　　　　　　093
　材料 Materials　　　　　　　095
　按照模板作画 Making Pressed Flower Picture According to Pattern　　095
　制作花窗玻璃式的凸面线条 Making Stained Glass Liked Outlines　　097
设计自己的图案 Design Own Pattern　　　　　　　098

多媒介
Multimedia　　　　　　　100
水晶花瓶插花 Arrangement in Crystal Vase　　　　　　　102
　材料 Materials　　　　　　　102
　制作水晶花瓶 Make Crystal Vase　　　　　　　104
　制作插花 Make Arrangement　　　　　　　104
蓝白陶瓷花瓶插花 Arrangement in Ceramic Vase　　　　　　　106
　材料 Materials　　　　　　　106
　制作陶瓷瓶 Make Ceramic Vase　　　　　　　107
　制作背景 Making Background　　　　　　　108
　制作插花 Making Arrangement　　　　　　　109
雨过天晴花瓶插花 Arrangement in Raindrop Vase　　　　　　　111

材料 Materials	111
制作雨过天晴花瓶 Making the Raindrop Vase	111
制作背景 Making Background	112
压连枝晶菊 Pressing the Daisies With Stems	114
制作插花 Making Arrangement	114

静物
Still-life 116

花篮 Flower Basket 118
春之庭院花篮 Spring Garden Basket 120
其他花篮 Other Flower Baskets 124

花瓶 Flower Vases 127

花盆 Flowerpots 131

幻想类
Fantasies 134

花仙子 Fairies 136
花仙子 Fairies 137
决定花仙子的种类 Deciding Type of The Fairy 138
花仙子的动态 Drawing Fairy in Action 141

丙烯泼画混合材料美人鱼
Acrylic Pouring Mixed Media Mermaid 144
材料 Materials 145
准备好工作台面 Prepare Work Surface 146
纸张 Paper 147
垫高 Padding 147
水平 Level 148
准备好泼的丙烯 Prepare Colors for Pour 148
泼画 Pouring 149
制作压花美人鱼 Making Pressed Flower Mermaid 149
制作美人鱼 1 Procedure for Making Mermaid One 151

制作美人鱼 2 Procedure for Making Mermaid Two 156

美人鱼另一例 Another Example of Mermaids 159

后记
Postface 160

01

压花艺术创意和技巧

Pressed Flower Art Ideas and Techniques

我在初、中、高级压花艺术书籍中教授了压花方法，基本美术构图理论，和制作各种物件的步骤。这本书志在唤起你灵感的同时让你掌握更多压花艺术的制作方法与技巧。令你在创作美丽的压花艺术大道上前进。

I have taught pressing methods, the basics of art composition, and processes of making many works in my beginner, intermediate, and advanced pressed flower art books. This book is for you to be inspired by creative ideas and to master more pressed flower art methods and techniques. You are on your way to create beautiful pressed flower art!

灵感来自日常生活。压花艺术与其他任何艺术形式一样，可以让您以视觉方式来表达。艺术赋予我们的灵魂以养分。一个人不需要艺术就能生存，但是肯定需要艺术才能生活。

Inspiration is from everyday living. Pressed flower art is just like any other art form that allows you to express visually. Art is nourishment for our soul. One does not need art to survive but certainly needs art to live.

　　您在自己的生活中是否有过一些瞬间，让您感动得无法用言语形容？您是否有创作某些东西的冲动，但又不知道该如何开始或怎么制作？还是您压了很多花但不知道该如何巧妙地利用？这本书将为您提供思路，让您探索并教您如何利用压花获得艺术效果的方法和技巧。这里，花朵和叶子不再仅仅是花和叶。您可以自由地使用花朵和叶子作为美术媒介创作出属于你的艺术品。

Have you had the urge to create something but don't know where to start or how to do it? Or you simply have pressed a lot but don't know what to do with your flowers? This book will give you ideas to explore and will teach you methods and techniques on how to achieve artistic effects with pressed flowers. There, flowers and leaves are not in their purest form anymore. You can use flowers and leaves as a medium to freely create your own unique piece of art.

压花艺术欣赏
Pressed Flower Art Appreciation

冬天是恢复和准备的季节。

——保罗·塞鲁

Winter is a season of recovery and preparation.

——Paul Theroux

静 物 创 意 压 花 艺 术

想看见花的人总会看得见。

——马蒂斯

There are always flowers for those who want to see them.

——Paul Theroux

我在森林里迷失但找到了灵魂。

——约翰·缪尔

And into the forest I go to lose my mind and find my soul.

——John Muir

光是画中最重要的一员。

——莫奈

Light is the most important person in the picture.

——Claude Monet

02

形状
Form

许多人在开始自己做压花艺术设计时因为觉得以前没有经过适当的艺术培训而裹足不前。我想您完全不需要担心这个问题。我们可以从日常生活中看到的一些东西开始。

Many people have problems getting started on their own designs because they feel that they did not have proper art training before. I think one does not need to worry about it. We can start from some simple ideas that we see from everyday life.

几何形状
Geometric Shape

我相信你每天都会看到盘子和碗。盘子和碗有三维立体形状（见《压花艺术（中级）》关于基本设计元素——形状）。我们通过俯视得到圆形来简化复杂的三维立体形状。让我们先从简单的圆形开始制作压花设计。

I am sure that you see plates and bowls every day. Plates and bowls have three dimensional forms (see book 2 about basic design element – form). We simplify the complicated three dimensional form by looking down to see circular shapes. Let's start from a simple circle shape with pressed flower design.

形状　Form

植物画
Cropped botanical

　　简单的圆形设计是采用一个圆形的纸框。

　　设计圆形的方式有很多种。其中一种方法是采用植物画设计。

A simple way to make a circle round shaped design is to cut a circular frameround mat.

There are endless ways to design a circle round shape. One way is to have a cropped botanical design.

1

先在卡纸上画上圆形，以便确定设计的框子范围。

Use the cut out to draw a circle on the backing cardstock in order to make sure that your design is within the frame.

2

在卡纸上贴好压花。

Attach pressed flowers to the backing cardstock.

3

覆膜，然后把纸框贴好完成。

Laminate the card with flowers. Attach the frame card onto the design to finish.

当你明白制作方法，你可以采用各种几何形状来设计压花。

Once you get this idea, you can do many designs with circles and other geometric shapes.

我们每天都会看到许多几何形状。我们看到长方形或方形的桌面。我们看到三角形的衣架。我们看到椭圆形的美式橄榄球。还有我们看到扇形的折扇。以几何形状为边界可以让您专注于装饰这种形状，从而提高您在压花艺术上的技能。

There are many geometric shapes we see everyday. We see rectangular table or square table tops. We see triangulartriangle shaped cloth hangers. We see oval shaped American footballs. And we see fan shaped folding hand fans. Using a geometric shape as the boundary would allow you to focus in decorating such shapes, hence sharpen your skills on pressed flower art.

我们可以使用长方形边缘为这张夏季婚礼请柬设计压花。

We can use a rectangle border design for a summer wedding invitation.

圆形也能用于比较复杂的设计。

The circular shape can be used in more complicated designs.

扇形通常用于中国风设计。

The fan shape is typically used in Chinese design styles.

装饰几何形状不限于把花材贴在形状的内部。可以将花朵固定在形状的边缘上。

Decorating a geometric shape does not limit to inside of the shape. One can anchor flowers on the border of the shape.

形状 Form

运用圆形制作一个完美的花环
Using Round Shape to Make a Perfect Wreath

使用圆形纸模是一种能制作出完美花环的简单方法。

Creating a wreath is easy with a circle shape guide.

材料 Materials

1. 小花（我用了角堇）
2. 叶材（我用了蕨）
3. 20cm×25cm或20cm×20cm水彩纸作为背景
4. 20cm×25cm或20cm×20cm白纸
5. 铅笔和尺子
6. 圆规或一个尺寸合适的盘子（15~16.5cm直径）

1. Small flowers (Viola, as depicted)
2. Greens (Lacy Fern, as depicted)
3. 8"×10" or 8"×8" watercolor paper as background
4. 8"×10" or 8"×8" white paper
5. Pencil and ruler
6. A pair of compasses or a circle shaped plates in desirable sizes (6"-6.5")

制作圆形状 Making the Round Shape Guide

1

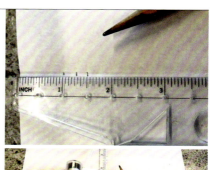

用和背景纸同样大小的普通纸，左右上下两边对折来确定中线。使用标尺在中心十字线上每厘米都标记一下。

With piece of paper the same size as the background, fold the paper in half, vertically and horizontally, to determine the center lines. Use a ruler to mark every 1/4" on the center cross lines.

2

如果有圆规，请画一个半径8.25cm的圆圈。或者可以使用大约16.5cm的小盘子。盘子的尺寸不需要精确，但不能太大，以至于边缘没有空间。

If we have a pair of compasses, draw a circle with 3.25" radius.

Or we can use a small plate about 6.5" across. The dimension of the plate does not need to be exact, but it should not be too big that you have no room on the edges.

3

把圆圈剪下来。剪的时候，里面的圆圈可以被剪坏但是外面的框子必须保持完整。用夹子把纸框固定在背景纸上。

Cut the circle out but keep the frame intact. It is fine to discard the inside circle since we will be using the frame. Use paper clips to hold the circle shape frame onto the background paper.

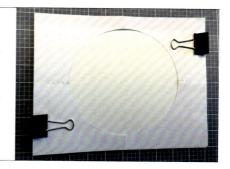

制作花环 Making the Wreath

在圆圈周围散布花。确保所有花朵都在圆圈内，边缘花朵就在圆圈边缘内。还要注意不要太贴近中心，避免将花朵排列成行。这将使设计过于僵硬。在每朵背后使用一点胶水将花朵固定。

Spread the flowers around the circle. Make sure all the flowers are inside of the circle and ones near the edge are just touching the guiding frame. Also, pay attention that we are not spreading too far into the center. Avoid lining the flowers into a pattern. That will make the composition too rigid. Apply a little glue on each base to hold the flower in place.

在花朵之间贴一些叶材。

Glue some leaves between flowers.

把设计放在数尺之外，调整花朵和叶子的位置来平衡设计。

Place the design a few feet away to view the composition in its entirety. Adjust flowers and leaves to balance the design.

变化步骤 Variation of the Steps

当然，您可以使用与我的《压花艺术（高级）》书中相同的技术，首先放置绿色藤蔓以形成花环的基部。现在使用圆形框更容易调整花环的形状。

Of course, one can use the same techniques as in my third book by placing the green vine first to form a base for the wreath. The circle guiding frame will help you to adjust the shape of the wreath.

变化花材 Variation of the materials

用于花环的材料可以非常灵活，可以使用花朵和叶子的任何组合。当然也可以只使用花。在这个例子中，我制作了一个只有花朵的凉爽夏季花环设计。

Materials used for a wreath can be very flexible. One can use any combinations of flowers and leaves. One could certainly use only just flowers. In this example, I explored made a cool colors with a summer wreath design using with flowers only.

变化框架 Variation of the Guide

如果担心花环中间，可以剪一个小圆形并使用纸胶带将其固定在背景上。在这个设计中，我首先铺开了铁线蕨，形成花环的底部。

If you worry about the center, you can use an inner guiding circle to create a perfect edge on the inside. Cut a small circleround shape and use double sided washi tape to hold it on the background. In this design, I spread out the maidenhair fern first to form the base of the wreath.

然后我铺上一些白晶菊。
添加几朵翠雀。

I then spread out some white daisies.
Added some delphinium.

将美女樱放在顶部或穿插，这样我们的设计就可以有点立体感。

Place some verbena either on the top or slot them behind other flowers to create depth in our design.

 绿色不能全部都在最下层。在花朵上面适量地增添一点小叶子，会使设计更灵活。

Not all greens are on the background. Add a few small leaves on top of flowers to make the design livelier.

在白色的花朵下面插入小叶子，令白色花材能从白色背景中凸显出来。

Tug a small leaf under the white flowers to make it stand out from the background.

 在上面增添一些卷藤丰富细节。
最后添加几朵蕾丝花完成。

Add some tendrils on the top for more details.
Finally, add a few Queen Anne's lace to finish.

球状
Spheres

我们做了几个圆形的平面设计。现在,我们要给圆一些体积感,让它们变成立体的。其中有一种方法是将光线投射到物体上,使其具有阴影。因此,物体看起来是立体的。在这幅画中,我们有两个球形花瓶。黑色因为本身色彩太暗,看不到阴影,但我们可以看到光的反射。我们可以在白色花瓶上看到阴影。

We have done a few projects with two dimensional circle shapes. Now we want to give the circles some volume so that they appear three dimensional. One way to do this is by casting light on the circles so they have shadows. Now the circles look three dimensional. We have two ball shaped vases in this picture. The black one is too dark to see shade gradients, but we can see light reflections. We can see shade gradients on the white vase.

形状 Form

材料 Materials

1. 水彩纸（20.5cm×25cm）
2. 双面贴
3. 花材：
    a. 黑色叶子（箭羽竹芋）
    b. 白色花瓣（飞燕草）
    c. 小玫瑰花蕾和叶子
    d. 葡萄风信子和叶子
    e. 咖啡色萱花叶子
    f. 原色叶脉

1. Watercolor paper (8"×10")
2. Double sided adhesive
3. Pressed material:
    a. Black leaf (rattlesnake plant)
    b. White petals (larkspur)
    c. Rosebuds with leaves
    d. Grape hyacinth with leaves
    e. Brown daylily leaves
    f. Natural color skeleton leaf

箭羽竹芋最好用微波压，会比较平整。

It is best to press the rattlesnake plant leaf with microwave. The leaves will be flatter.

制作叶脉 Making Skeleton Leaves

我将白兰叶放在一个带盖的大塑料盒中泡清水，在夏天将其放在外面7~10天。叶肉会变得很软。用清水把叶肉洗净，叶脉就出来了。用纸巾擦干叶脉，然后用厚书或任何压花板压。

Submerge magnolia leaves in a large plastic box with a lid and leave it outside for 7-10 days during the summer. The mesophyll of the leaf would become very soft. Isolate the leaf's veins by washing away the mesophyll with water. Dry the skeleton leaves with a paper towel and then press them with thick books or a flower press.

制作球形花瓶 Making the Ball Shaped Vases

1. 首先，我画两个花瓶（7cm和6cm直径）。你可以先画两个圆圈，然后修改底部和瓶口形成花瓶。

First, I drew two vases (3" and 2 3/8") diameters). You can draw 2 circles and then flatten the bottom and the top opening a little bit to make them vases.

用电脑反转图像，打印。或是用描图纸描了边缘之后，再用马克笔把边缘勾勒一遍，使反面也可以看清楚边缘。把双面贴贴在打印的白色面或是描图的正面。用剪刀把形状大略剪下来（边缘留一些白边）。

Flip the images with a computer and print. Alternatively, trace the edges of the vases with a images, use marker on the edges so you can see them clearly on the backside. Apply double sided adhesive on the back white side of the printed images or the front side of the trace. Cut the shapes loosely (not too close to the edge).

用黑色叶子把小花瓶贴满。用镊子掰出光斑的形状，贴好。

Glue black leaves onto the smaller vase.

Use tweezers to shape the light spot and glue to the vase.

将白色花瓣贴在浅色区域。使用镊子掰出叶脉的小块，以形成球形的阴影区域。将叶脉的深色叶肉残留部分用于较深的阴影。

Glue white petals on the light area. Use tweezers to nip small pieces of the skeleton leaf to form the shaded areas of the sphere. Use the darker meat portion of the skeleton leaf for the deeper shade.

把枯萎的萱草叶（或任何长脉叶）熨烫平整对于制作木质平台非常有用。

Dead daylily leaves (or any long vein leaves) ironed flat is suitable for making a wood platform.

6

把花瓶、木台和桌面线条贴好。

Glue the vases, wooden platform, and the table line onto the watercolor paper.

7

把玫瑰花蕾和葡萄风信子插瓶贴好。
用浅色叶脉贴在桌面的位置。中间淡深色的叶脉贴在阴影部分。

Arrange the rosebuds and grape hyacinths into the vases.

Glue light colored skeleton leaves onto the table space. Glue medium colored skeleton leaf pieces to form shadows.

长方盒状
Box Shape

我们将运用透视来制作一个窗边窗口花坛。这是另外一个可以获得立体感的方法。

We will try to make a window box in perspective point of view. This is another way to create a three dimensional 3-D effect of a shape.

材料 Materials

1. 水彩纸（15cm×20cm）
2. 双面贴
3. 花材：
   a. 牛蒡皮
   b. 小花、叶子和梗

1. Watercolor paper (6"×8")
2. Double sided adhesive
3. Soft pastel or pan pastel
   Pressed material:
   a. Burdock root skin
   b. Small flowers, leaves, and stems

很多根茎类的蔬果都适合于制作木材，例如，土豆、防风草（白胡萝卜）和木薯。把削下的皮用任何压花器或是厚书压。

Many root vegetable skins can be used to depict wood. For example, potato, parsnip, and cassava, to name a few. Just save the peels and press them in a flower press press or a thick book.

形状 Form

制作窗边花坛 Making the Window Box

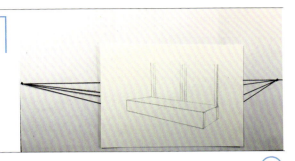

首先我们画一个窗边窗口花坛的两点透视图。

First, we draw a two point perspective view of the window box.

我在窗边窗口和花坛周围均匀地涂上背景墙体的颜色。

I blended background wall colors onto the areas surrounding the window and window box.

用描图纸把花坛的可见形状描下来。靠墙边缘和左边到时候会被花遮住，所以不用描。用马克笔把顶点都标注好，这样从反面也可以很清楚地看到。

Use tracing paper to trace the visible shape of the window box. The edge against the wall and the left side will be covered by flowers, we do not need to trace. Use a marker to mark the points that form the shape so we will be able to see from the back side easily.

把花坛两块板剪开。描的图正面覆盖好双面贴。

Cut the join to separate the box walls. Apply double sided adhesive to the front side of the trace.

5

把牛蒡皮贴好。然后把形状剪好。

Glue burdock skin onto the trace. Trim the shapes.

6

把两边木板贴在背景上固定贴好。另外剪两条细条牛蒡皮。在右边竖贴，以显示出木板的厚度。

Glue the two sides onto the background. Glue two thin strips of the burdock skin to show thickness of the boards on the right side.

7

用小梗子和叶子拉出植物的高度。用叶子把花坛的表面白色空间全部遮住。我使用的是天葵。大爱其线条和细致的叶子。

Use small stems and leaves on the window box to define the height of the plants. Cover the white spaces on the surface of the box. I used semiaquilegia in this picture. I love the lines and delicate leaves.

8

把小花安排好，完成。

Arrange small flowers to complete the picture.

混合形状
Combination of Shapes

　　我们可以把两个或更多个几何形状组合起来形成更复杂的形状。我们经常会在日常生活中看到它们。当然，我们也可以使用这些形状来制作压花艺术设计。

One can combine two or more geometric shapes to form a more complicated shape. We can see them in our daily lives. Of course, we can also use these shapes to make pressed flower art designs.

　　这里有一些花瓶的形状。

Here are some flower vase shapes.

　　可以装饰每个瓶子，塑封，然后制作一张卡片。

We can decorate each vase, laminate it and then make a card.

　　我在《压花艺术（初级）》书中详细介绍过塑封和卡片制作。我这里就不再细说了。

Use the process as described in my first book to laminate the card.

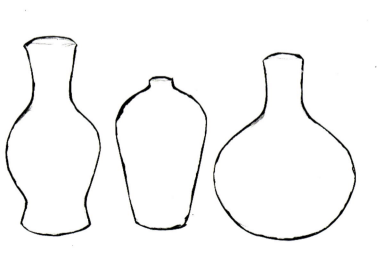

组合形态
Combined Form

组合大丽花 Assemble Dahlia

1. 最好是使用色调深浅不同的同种花或叶子。这样成品花瓶外观会比较顺滑。我用深浅不同的几种蓝色和白色的绣球花。

It is the best getting the same flower or leaf with different colors. The finished vase would have a cohesive appearance. I use hydrangea in several shades of blue and white.

2. 像往常一样,我通常会对物体进行绘图研究。通过学习,我可以更多地了解物体的光影。首先,我决定光源来自右上方。

As usual, I do a drawing study of the object. Through study, I learn more about the light and shade of the object. First, I have decided that the light is coming from upper right.

3. 使用马克笔使花瓶的轮廓更清晰。在花瓶图案的上面粘贴双面胶。

Using a marker to make the outline of the vase clear. Adhere double sided adhesive on top of the vase pattern.

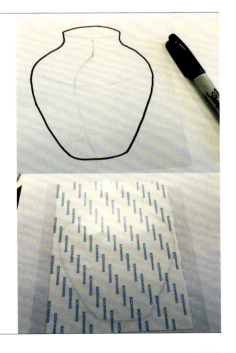

把形状大致剪下来（不要剪得太贴近边缘）。

按照绘图的光影深浅，贴上绣球花瓣。在交错的地方，用镊子稍微把绣球的边缘捏掉一点，使其有不规则的口。

Glue the hydrangea petals according to the light and shade of the drawing. In the overlapping portions place, use a pair of tweezers to slightly overlapping the edge of the hydrangea to make it have an irregular shape.

用绣球花瓣把花瓶贴满。

可以使用中性玻璃胶粘贴绣球花瓣来细调花瓶的光影。

Glue the hydrangea petals to cover the entire vase.

You can further adjust the colors and make the shape more three dimensional by gluing some petals to adjust the places you see that need to be fixed. Make sure you use acid free silicone glue so the petals will not have spots.

翻过来把花瓶形状剪出。

Flip over to cut the vase shape out by following the edge you drew.

形状 Form

7

把花瓣按照大小分好。

Sort the flower petals according to sizes.

正面 Front view

1

剪一片2.3cm 的白纸，贴上双面贴。也可以不用白纸而使用单面贴。

把4个角修圆。揭开离型纸。

Cut a piece of 1" (2.3 cm) white paper. Cover it with double sided adhesive. Alternatively, you can use single sided adhesive (white label) without white paper.

Round off the edges. Peel off the adhesive protection paper.

2

把8片大花瓣贴在白纸上。然后再贴一圈8片花瓣。贴第二大花瓣一圈。如果你的花花瓣很密，可以再贴第4圈和第5圈。

Glue eight large petals onto the white circle. Repeat again with another eight petals.

Glue eight second large petals for the third round. If your flower has more petals, glue the fourth and fifth rounds.

3

把白纸贴上双面贴，或是使用单面贴。剪2.5cm的圆边三角型。把最小花瓣对着一个尖角贴。

Place a piece of double- sided adhesive onto a piece of white paper or use singled sided adhesive instead. Cut a 1" rounded sided triangle. Use small center pieces and point all the tips to one side of the triangle.

4

用第二小花瓣竖着围着尖角贴一圈。翻过来把多余的竖立花瓣尾按照形状剪掉。

Glue some second small petals vertically toward the tip of the triangle. Flip over and trim off the vertical petal ends according to the shape.

5

把花心贴在之前组合的花中心完成。

Place the center onto the partially assembled flower to finish.

形状 Form

1

侧背面 Side Back View

先用4片中型大小花瓣（背面朝上）形成一个扇型。然后在上面交错再用4片大型花瓣（背面朝上）形成一个扇形。

Arrange four medium petals like a fan (back side of the petals face up). Then arrange four large petals (back side of the petals face up) to form another fan but with the under petals peeking out.

侧面 Side View

1

4个大花瓣形成一个扇形，然后排列3个中花瓣以在顶部形成一个扇形，花瓣交错。

Four large petals to form a fan shape, then arrange three medium sized petals to form a fan shape on the top with the back row of petals peeking out.

2

再用4片中型花瓣（背面朝上）形成扇形。后面两排的花瓣要探头出来。

Arrange four medium petals (back side facing up) to form another fan shape with back two rows of petals peeking out.

3

4~5个大花瓣（背面朝上）在前面三行花瓣上再形成一个扇形，让前面三行花瓣探头出来。

On top of three rows. Let the back rows of petals peek out.

041

5 选择浅色，两种中间颜色和一个深色为墙体。选择中间颜色的棕色和很深的颜色为桌面颜色。

Select a light tone, two shades of medium tone, and very dark leaves/flowers for the background wall. Select a mid-tone brown and a very dark material for the table.

6

把背景颜色简单铺一下查看色彩是否搭配。如果不够搭配则需要选择别的背景颜色。

Tease the color mass to determine if the colors go together. Select other leaves if the background color mix does not work.

首先将墙壁粘上。从黑暗到明亮排列叶子。如果你有深色典具纸，可以使深色典具纸覆盖左边的2/3墙体使颜色过渡平顺。注意：用湿毛笔在典具纸的边缘画上线条，撕出毛边。这样色彩过渡会更自然。然后贴桌子。花瓶将置于桌子的阴暗面，因此较浅的颜色将占据右侧表面的约1/3。

Glue the wall first. Arrange leaves from dark to light. Cover left 2/3 of the wall with dark tengucho paper if you have it to make the color transition smoother. Note: use a wet brush to draw a line around the edge of the tengucho paper and then tear the edge so it is not a sharp cut. This would make the color transition look more natural. Then work on the table. The vase will stand on the dark side of the table so the lighter color will occupy about 1/3 of the right side surface on the right.

然后将花瓶和大丽花粘上。将配花插入花瓶，使瓶花更饱满。我用的是绣球花。请注意，光线来自右上角。因此，前面和右前的大丽花比后面和左边的颜色要浅。

We then work on placing the vase and dahlia flowers. Insert a companion flower into the arrangement to make it fuller. I have used hydrangeas. Note the light is coming from upper right. Therefore, the flowers in the front and slightly right flowers are lighter in color compared to the back flowers in the back and left flowers.

形状 Form

有机形状
Organic Shapes

形状不一定是无机的。我们也可以在压花设计的时候运用有机形状。

Shape does not need to be geometric. We can use organic shapes in pressed flower designs.

轮廓方法 Outline Method

1 宠物爱好者们，大家可以拍摄自己的宠物照片，然后描出轮廓。尝试简化轮廓（不捕捉每个小凹凸）。

There are many pet lovers. You can take a picture of your pet and then use it for the outline. Try to simplify the outline (don't capture every little ·bump). You can trace my drawing of cat if you want to.

2 把形状剪下来，然后用形状作为模板在背景上画下来。用刻刀把形状刻下来。用黑马克笔把形状的边缘描黑。

Cut the traced shape out and use the shape as a guide to draw on the background. Carve the shape out. Use the black marker to paint the rim of cutting.

3

在另一张背景卡纸上设计好压花。覆膜。把压花设计贴在背景后面，完成。

Apply flowers on a separate background card stock. Laminate. Glue the pressed flower design behind the background to finish.

043

这里是使用我画的猫图片制作的压花卡片。

Here we have another card with my cat drawing.

拼图方法 Puzzle Methods

1. 我们可以绘制一些基本的具有识别力的图案。我通常把我绘制的图扫描，然后反转变成反方向的图。打印一张正面图和几张反转图。把各个需要不同颜色的区域在反转图中大致剪出。然后把双面贴粘在每个区域的背面。

We can draw some basic drawings that distinguish patterns. I usually scan my drawings and flip the images with software. Print out one normal image and a few flipped images. Roughly cut out different parts of the images that is intended for different colors using the reversed images. Then apply double sided adhesive on the back of each area.

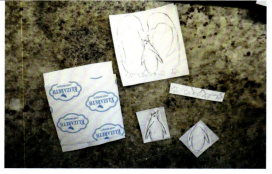

2. 贴上花瓣或叶子作为颜色。

Glue petals or leaves for the color.

形状 Form

3

反转，按照图形剪出形状。
把所有不同颜色的区域好像拼图一样拼在正面图上。把整个图剪出来。

Flip to the back side and cut out the shape for the color.

Fit all the different color parts like a puzzle onto the normal image. Cut out the entire image.

4

在水彩纸上均匀地涂上不同的蓝色。撕一张纸放在下端阻挡颜色。

Blend different blues onto a piece of water color background paper. Tear a piece of paper as template for the bottom.

5

在背景上喷点水，晾干。可以用吹风机或热风枪来加速干燥过程。

Spray water onto the background and let it dry. We can speed up the drying process with a hair dryer or a hot gun.

6

贴一些白色香雪球花作为雪花在背景上增加细节。

Glue some white alyssum florets as snowflakes on the background for details.

045

形状 Form

有机形态之真实感的玫瑰
Organic Form – Realistic Roses

花卉都属于有机形态，有各种各样的形状。我选择玫瑰来讲有机形态，因为玫瑰有很多种类和形状。单瓣和两层花瓣的可以整朵压。在中国，又分蔷薇、玫瑰和月季。本篇中提到的玫瑰，是大型月季，以普通称法泛称为玫瑰。本篇集中在大型重瓣玫瑰特别是花束中使用的长柄玫瑰。这些玫瑰有很多花瓣包卷形成立体的形状。

Flowers are in all kinds of organic forms. I want to pick roses as an example for organic forms because there are many varieties of roses with different shapes. The single petals of semi double ones can be pressed whole. This section, however, concentrates on the large double roses - especially the long stem roses often used in bouquets. These roses have many petals that wrap around to form a three dimensional rose.

通常花束中使用的玫瑰是初开，而不是像我们在花园里看到那种完全开放的。这种初开的玫瑰最能体现出其美和清新。

Usually, roses used in a bouquet are not fully opened like you have seen in gardens. The slightly opened rose is at its best stage in turns of beauty and freshness.

当我们压花时，我们把玫瑰花瓣拆开分别压。如何把压好的平面花瓣组合成玫瑰，是比较困难的事情。

When we press roses, we press individual petals. How to assemble the flat petals back to form a rose ishas always been a difficult task.

玫瑰组合 Assemble Roses

凸显玫瑰美丽模样的组合方法有很多。建议先学一种组合方法，练习熟了再学下面一种。

There are many ways to assemble roses to show case their beautiful characteristics. I suggest you start with one method, practice it. Move to the next one after you have mastered the first method.

首先，我们把玫瑰花瓣从保管袋中取出，放在室内1~2小时，让花瓣吸收空气中的湿气。这样花瓣变软，不会脆得不能制作。如果空气太干燥，可以用毛巾喷了水，拧干。放在微波炉用盘子里，微波20秒。然后上面盖上干毛巾，把花瓣放上。等10~20分钟让花瓣吸收蒸汽变软。

First, leave the petals outside in room temperature for 1 to 2 hours so they absorb some moisture to become softer. If the air is too dry, try this. Wet a small towel with warm water slightly and squeeze out excessive water. Place the damp towel in a microwave safe dish and microwave for 20 seconds. Place a dry towel on top of the damp one and then place the petals on the top. Wait for 10-20 minutes for the petals to get softer.

形状 Form

　　我们面对的一个问题是玫瑰花瓣比较大。而我们通常制作20cm×25cm或28cm×35cm的画。那我们怎么对付7cm或以上的花瓣？

One of the problems we face is that flower petals are big. And most of pictures we want to make are 8"×10" or 11"×14". What are we going to do with 3" or larger petals?

　　在组合玫瑰之前，必须先提一下胶。玫瑰花瓣对胶比较敏感，尤其是淡色玫瑰更甚。最好用中性玻璃胶或是日本的花胶。

Before assembling roses, we need to discuss about the glue. Rose petals are sensitive to glue, especially the lighter colors. It is best to use acid free silicone glue or the Japanese resin glue.

049

有角度的玫瑰组合 Assemble Roses Angle View

1. 取一7cm或以上大花瓣，仔细看一下。花瓣朝内侧（朝花芯的一边，通常比较深色，也不会发亮。当花盛开的时候，这边是我们看到的。把花瓣的内侧朝上，剪成4片如图样。

Take a large petal (3" or more). Take a look of the petal. The petal face inside (facing the center) is usually darker and mat. When the rose open up, this is the side that we see. Now, have this side face up and cut it into four pieces as shown.

2. 取上面剪出来的小片，折两下如图。

Take a cut piece from the upper petal, fold as shown in the picture.

3

把上面的小片粘起来，成三角形。上面留一个非常小的孔。这个是做玫瑰中间还没大开的花瓣。下面两片翻过来，把花瓣外侧折小角过来如图。

Glue the top folded petals to form a triangle with a tiny opening on the top. This is making the un-opened center. Flip the bottom cuts and fold the outside rims as shown.

4

把两片折过的花瓣互相交错地贴在小三角上。

Glue the two folded lower cuts onto the little center cross overlapping as shown.

5

把两边向后折，贴好。

Fold the two sides back and glue.

6

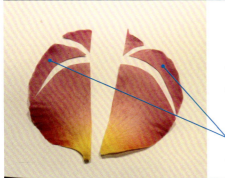

取一片大花瓣，剪成图中样子。

Take a large petal and cut as shown.

这两片是第8步需要用的。

These are what we will be using in step 8.

7 第一片大花瓣剪出来最后一片，再把旁边修圆。

The last piece from the first large petal needs to be trimmed as shown.

把两片第6步剪下来的花瓣交叉贴在前面。注意两片的上面尖处要靠近玫瑰组合的花芯顶端外侧。然后把第7步剪下来的花瓣贴在后面，对齐交叉花瓣的顶端。 8

Glue the two cut outs from step six cross over on the front. Pay attention to the top of these two petals. They should be very close to the top of the assembled rose. Place the cut out from step seven to the back of the rose assembled so far. Make sure the top is connecting to the crossing petal tips.

 把交叉花瓣下边剪掉。

Cut the bottom along the lower rim of the crossing petals.

第6步剪下来的花瓣，再剪成图示的样子。箭头所指就是我们需要的。这步之后我们需要的花瓣都是剪成大致这个样子。花瓣的边缘不是圆的，而是稍微有点尖。但是不要太尖。 10

Cut the larger lower pieces from step 6 into shapes like in the picture. These are the ones we will use. Basically, all of the outer layers from the next step and on will need petals like these. The rim is not round but a little bit pointy. The slope should not be very sharp.

把两片花瓣粘在玫瑰组合的下边。

Glue the two petals under the assembled rose.

取一片大花瓣,剪成4份,然后每份剪成刚才第10步那样的花瓣。

Take a large petal and cut out smaller pieces like the ones we had in step 10. Repeat with the other pieces so we have 4 cut outs per one large petal.

把4片花瓣粘在组合的玫瑰下边一圈。

Glue those 4 pieces under and around the assembled rose.

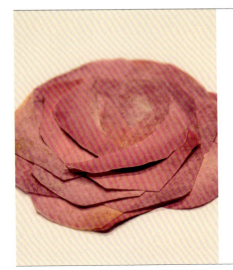

当我们粘外圈的时候,一片玫瑰花瓣只能剪出两片花瓣,因为我们需要更大片的花瓣。如果有小花瓣,这时就可以不需要剪,直接用。我们需要在组合的花下边3/4处多贴两圈花瓣。上边不需要贴。

As we move to the outer layers, we will just get two pieces per petal since we need larger pieces. We will also need more petals per layer. If we have smaller petals, it is fine to use them as is. We need to glue about two more layers after step 13 but only on the lower 3/4 of the flower. There is no need to glue more petals on the top.

15
如果看到有缺口，可以用花瓣填补起来。

Fill up the gaps as we see.

16
组合完成。这朵花的花瓣比较圆。如果希望花比较像前面图示的，可以把花瓣折一下。不是每种玫瑰开起来都会翻瓣。你按照需要决定是折还是不折。

Finished rose. This rose has round open petals. If we want the rose to look more like the picture with rolled edge petals, you can fold some outer layers of petals. Not all roses open with edges rolled. You need to look at the rose you are working with to decide.

正面的玫瑰组合 Assemble Roses Front View

形状 Form

1 如果你仔细看玫瑰花瓣，它上下有深浅。还有它的正反面色彩和反光度也有区别。

If you look at the rose petal, there is darker portion and lighter portion. Plus, the color and glossiness between front and back of the petal is a little different.

2 取一片花瓣，剪出至少两种不同色彩的地区。然后剪出左图的模样。先剪出一个圆形，然后再顺着边剪比较容易。

Take a petal and cut at least two different color sections out. Trim them to small curves like shown in the picture. It is easier to first cut out a circle and then cut around the edge.

3

把小片弧形贴在花瓣的背面。

Glue small curved cuts onto the back side of a petal.

4

继续贴弧形小片。

Continue to glue more curved cuts.

5

转圈贴弧形小片形成玫瑰花心。

Glue more curved cuts to form the center of the rose.

把玫瑰花心剪下来。

Cut the center out.

剪一片大花瓣。

Cut a large petal like this.

把剪出来的碎片修剪成花瓣。

Trim the cuts into rose petals.

把"花瓣"的根部剪掉。

Cut the tip of the "petals" off.

把剪好的花瓣转着圈贴在组合的花心后面。因为花瓣越接近花心就包得越紧所以每瓣边缘露出少许。

Glue the pieces one at a time around the assembled center. Show small edges since petals are tight closer the center.

11

继续增加层数，但是用花瓣背面。

Add more layers but with the back of the petals.

12

下一轮一片大花瓣剪成4片，然后转圈贴。
再下一圈贴背面花瓣。

Next round, cut a large petal into four parts and then arrange circling to glue the petals. Repeat with the back side of the petals.

13

把整片花瓣裁短，外围一圈用整片花瓣来贴。

Trim the petal short and use the entire petal for the outside round of petals.

14

完成的正面玫瑰。

Finished front view rose.

半开和全开的玫瑰组合 Assemble Half and Full Open Roses

1

把一片大花瓣剪成4片。

Cut a large petal into four pieces.

2

把花瓣剪出来。需要15片这样剪出来的花瓣。其中5片要比其他的略小。注意，只是略小，不要小很多。

Trim these petals. We will need fifteen of such cut outs. Have five of them slightly smaller than the rest. Notice, It is just slightly smaller, not too much smaller.

3

把5片大花瓣贴在一张圆纸上。

Glue five larger petals on a circular paper as shown.

4

换一个角度，再把另外5片较大花瓣往中间一点粘。

Alternate positions and glue another 5 larger petals slightly more towards the center.

5

再粘略小的花瓣。要从中心粘。注意角度要和其他两层错开。

Add one more layer of five petals. Make sure these petals are slightly smaller than the last two layers. Glue them to cover the center. Make sure they are not lined up with either of the layers.

6

这时粘上玫瑰花芯，就是全开的玫瑰。

For a fully opened rose, add the center.

7 制作半开的玫瑰，把1片小花瓣折叠成三角，上面有一个小的口。这是还没开的花芯。

For half opened rose, fold a small petal to form a triangle with a tiny opening on the top. This is the unopened center.

8 再折2片小花瓣。然后按照中心线对折。

Fold two small petals as shown. Then fold it again along the center line.

9 把折好的花瓣包在花芯上，贴好。然后同样再贴右边那片。

Glue the folded petals as shown and sandwich the center. Repeat on the right side as well.

10 取1片中型花瓣，剪开。一边要比较大。把顶端折一下。不要折太多。

Cut a medium size petal in half (one side slightly larger) and then fold the top as shown. The fold should be small.

形状 Form

11

先把小的那片粘好。然后再粘大的那片。

Glue the smaller one first and then the larger one on the top.

12

粘1片中型大小的花瓣在后边。注意前面折叠的顶端要和后边那片花瓣的顶端对齐。

Glue a medium size petal on the back. Notice the tip of the front folded petal need to line up with the back petal edge.

13

把花修剪一下。如果你开始不知道怎么下剪，可以先把两边和下边用直线剪出。然后再慢慢修剪。

Trim the rose assembly to a deep bowl shape. If you have trouble at first, cut straight lines on two sides and bottom first. Then slowly trim.

14

把微开的中心部分粘到第5步完成的花瓣上。完成半开玫瑰。

Glue the assembled center to the rose petal assembled from step 5. Now we have a half opened rose.

061

玫瑰花束制作 Making Rose Bouquets

　　玫瑰组合完成之后，我们看一下花束。花束的顶端和旁边，我们看过去都是不同的角度（蓝色标出来的地区）。只有黄色这区面对我们的，才可能看到正面的玫瑰。有时其实也是不同的角度。再看绿色材料。一般都在下边（绿色标示）。只有采用了直立长枝的材料，才会在上边冒出头。绿色材料如常春藤、柑桔枝，特别是玫瑰叶子，是不可能在上面冒出来的。

Now that we have roses assembled, let's start making the bouquet.

Take a look of a bouquet. The top and the side roses are viewed at an angle as marked in blue. Only the middle ones are facing us as marked in yellow. Sometimes even the middle ones are viewed at an angle. Therefore, if you only learn to assemble rose one way, learn the angled view. Notice that if greens are used, most of the greens are showing on the bottom of the bouquet as marked in green. Greens are peaking on the top only if it is on a long stem of vertical material such as Irish bells. Greens such as ivy, many citrus leaves, especially rose leaves will not peek out on the top.

关于背景

如果准备运用枝干和缎带，还准备真空密封这幅画，背景质地必须要软和富有弹性。如果使用纸张，最好先仔细搓揉一下，再用手小心地把它展平。这样可以把硬挺的纤维变软，让作品背景在真空密封时不会出现皱纹。

About background

If we are planning to have stems and ribbons for the bouquet, and are also planning to vacuum seal the picture, then the background must be soft and slightly elastic. If we are using paper, it is much better to knead the paper and then flatten carefully with our hand in order to break down the stiffness of the material to avoid creases when vacuum sealing the picture.

关于缎带

很多人都对花束的枝和缎带制作感觉困难。这里我谈两种做法。大家应该对制作一个圆形的花设计没有问题。最主要是要留位置给枝和缎带。我通常把所有的元素都先准备好才着手粘贴花。这样我到后面制作的时候就不会发生问题。

About Stems and Ribbon

Most of people have trouble with the stems and the ribbons. I will talk about two ways to work with stems and ribbons. I am sure everyone can arrange the flowers to form a round bouquet. Just make sure we position the round flowers well so you have room for stems. I usually make all the parts first before I glue the flowers down for the design so I will not have issues later.

1 准备好枝子。这些都是剖半压制的。如果太粗厚，可以用砂纸磨一下。我通常把末端剪一个角度。你也可以把它们统统剪成一个长度。

Gather the stems. These were pressed sliced in half and pressed. If they are still very thick, you can trim off the sides slightly or sand them down a little bit. I usually cut the bottoms in an angle so they look better. You can trim them to the same length if you want.

2　剪一段缎带，贴上双面贴。然后把枝子贴在一半处（缎带1.5寸宽）。注意不要重叠。

Cut a piece of ribbon. Apply double sided adhesive. Glue the stems about half way. (Ribbon is 1.5" wide). Make sure they do not overlap.

3　把双面贴贴在另外两段缎带上，然后贴在枝子的上端。可以这样交叉贴，或是顺一个方向。看拿到的花束缎带绑法决定。

Apply double sided adhesive and glue two pieces of ribbons on the top. We can either angle them the same way of cross them depending on how the original bouquet was tied.

4　把两边剪齐，然后把两边折在下边。注意：两边需要剪到刚刚折进去，而不交叉。不要再增加更多的厚度。

Trim the sides and fold the sides in. Note: trim the sides so the folds were just in and not overlapping to cause more thickness.

5　现在我们的缎带和枝子成为设计的一个元素。我们可以看到这个元素的上半段是没有枝裹在里面的。不那么厚的时候比较容易在上面摆放花叶。

Now we can use the stem and ribbon as a unit for your picture. We can see that the top half of this unit actually has no stems in there. It will be much easier to place flowers and leaves onto it where it is not too thick.

形状 Form

现在我们看看另外一种做法。Now we will examine another method.

1

我们也可以利用不干胶，放在缎带的中间。

We can also use an adhesive dot runner. Apply in the middle for about 1/2"-3/4".

2

把缎带的两头1/3处剪45°角。然后把两头贴向中间。

Trim 1/3 off at edges at an angle. Glue the two ends to the center.

3

把中间折一下。这是正面看过去。

Fold the center in a little. This is the front view.

4

剪一小段缎带，放上不干胶。

Apply the dot runner to a small piece of ribbon.

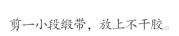

贴在刚才的缎带中间。

Glue it to the center.

包过来。

Fold it over.

现在看看，它平整让我们容易做压花设计。

It is nice and flat for us the work with.

8

再贴上两根飘带。

Glue two pieces under.

现在我们有一个扁而平整的蝴蝶结。系的蝴蝶结对于压花设计来说是太厚重了。

Now you have a nice flat bow. Tying the bow will make the ribbon too bulky for pressed flower design.

形状 Form

10

把蝴蝶结塞在花束的花下边。

Tug the bow under the flower design.

11

和刚才一样，粘一些枝在丝带上。

Like before, glue the stems onto the ribbon.

12

修剪两边，然后把两边往内折。

Trim and fold over the ribbon.

现在我们已经组合了玫瑰，茎和缎带形成一个单元，设计花束变得更容易。还记得我们之前看过的真实花束吗？我们按照真实的花束来安排鲜花。

首先安排玫瑰，这样你就可以确定位置，以确保花束不会不平衡。在安排玫瑰的时候记得要预留足够的空间给茎。放置茎和缎带。只有花束中心的地方我们才能看到玫瑰的正面。一旦安排好玫瑰之后，中间用一点胶水将玫瑰粘好。

Bouquet Design:

Now that we have the roses assembled and the stem, and ribbon forming a unit, designing the bouquet becomes easier. Remember how a real bouquet we have seen earlier. We arrange the flowers using the earlier bouquet example as a reference.

Arrange the roses first so you can determine the positions to make sure the bouquet will not be lopsided. Make sure that we have enough room for the stems when arranging the roses. Place the stems and ribbon. Only the very center ones will we would see a full face of rose. Once the positions look right, glue the roses with a little glue on the center to hold themit down.

　　请注意，在真正的花束中，我们不会看到所有的玫瑰都在其他辅花的上面。因此，我们不能只是简单地把其他花材填补在玫瑰下面，而是要让其他花材也稍微压过玫瑰。我们必须留意透视角度。花需要从中间压往四周，我用黄色箭头标示了方向。

　　将茎插到缎带蝴蝶结下。您可以在缎带顶部放一些小花茎或叶子，使其更加生动。

Note that in a real bouquet, we would not see roses all above the other flowers. Therefore, we should not just tug the other flowers under roses. Instead, we should allow other flowers to the overlapping roses a little as well. Pay attention to the perspective. The flowers over lapping directions are as indicated in the yellow arrows.

Tug the stems under the ribbon bow. You can allow some small flower stem or leaves to be on top of the ribbon to make it livelier.

形状 Form

玫瑰花园制作 Making Rose Garden Picture

有时我们在保存一束花时并不希望制作一个花束，而是一幅能够装饰家庭的画。玫瑰花园这种设计很受欢迎。如果是客人委托我们制作遮掩的设计，我们必须先和客人沟通好，明白我们需要加入一些花束以外的玫瑰叶子。我的经验是只要花束里面的材料必须占画面的大部分，客人不介意我们加入叶子但加入额外的叶子无碍。

Sometimes when we press a bouquet you do not want a picture of a bouquet but an artwork that you can use for home decoration. A garden design is very well liked by many. To make such a design, we need to know that we will have to add rose leaves to the design that are not found in the original bouquet. My experience is that as long as the original flowers are the main parts of the design adding additional leaves would be fine.

1 粘一些玫瑰叶子在白色背景（8×10寸）上。一边的上方留白。

Glue some rose leaves on the background (8"×10"). Leave one upper corner open for some space.

2 上面盖一层蕾丝纸或其他薄半透明纸。但如果纸太透明，可以使用两层。

Place the lace paper (or any sheer paper) over on the leave stop. The idea is to create some depth of field the rose garden. If you are using sanwa, maybe consider to use two layers since sanwa is very sheer.

3 粘一些叶子，反面朝上。

Glue a few rose leaves backside up.

4 粘一些玫瑰叶子，正面朝上。正反面色彩的差异，给人一个视觉上丰满花园的感觉。

Glue some rose leaves front side up. The color difference between the back and the front will create more depth and visual appeal for a lush garden.

5

摆放玫瑰花。有些花需要一些重叠，这样才比较有立体感。

Arrange the roses on the top. Some overlap is necessary to create depth.

6

把叶子插在前面和后面的玫瑰花中间。

Place leaves under the front rose.

7

高出来的玫瑰需要有枝。

Add a stem for the rose peeking out of the bush.

8

玫瑰枝上要有小叶子。整体看一下，需要多叶子可以再补。

Add leaves to the rose stem. Add more leaves where necessary.

用长剪刀修剪四边。签名完成。

Trim the edges with a pair of sharp long scissors. Sign to finish.

　　我们也可以用从《压花艺术（高级）》书中学到的制作鸟的方法，来作一幅玫瑰园中的对话。

We can also use the same technique we have learned in my Advanced Pressed Flower Art book for birds to make a rose garden with singing birds.

03

图案

Patterns

图案是通过重复或呼应视觉元素来形成的。这种重复或呼应传达了平衡、和谐、对比、节奏或动感。

Pattern is made by repeating or echoing the visual elements of an art work. The repeating or echoing of the elements will communicate a sense of balance, harmony, contrast, rhythm, and or movement.

　　我们可以将图案式设计应用在整个画面或画面的一部分。首先，我将讨论一下图案的类型和运用此等图案的压花作品。然后，我们将讨论一些创新的图案使用方式。

We can apply patterns for the entire picture or a portion of the picture. First, I will discuss some types of patterns and sample pressed flower picture. Then we will discuss some more creative ways of using patterns.

压花作为设计元素
Pressed Flowers as Design Elements

压花非常适合作为图案元素。在一块手绘丝绸上安排这种简单花朵，花朵的上下跳跃营造出这幅画的韵律。

Pressed flowers are perfect as pattern elements. For this simple pressed flowers over a piece of hand painted silk, the flowers running up and down creates the rhythm of this picture.

对称
Symmetry

对称在生物界中无所不在。

Symmetry is pervasive in living things.

旋转图案
Spiral Pattern

螺旋是一种弯曲的图案,其聚焦在中心点上,并且围绕它旋转的一系列圆形形状。这是一幅抽象画"太阳"。

A spiral is a curved pattern that focuses on a center point and a series of circular shapes that revolve around it. This is an abstract picture "El Sol".

流水图案
Flow Pattern

《冰雪花》流水图案让压花充满动感。

The flowing pattern can be use in pressed flower art design to show movement. "Ice and Snow Flowers"

波浪，沙丘图案
Waves, Dune Pattern

当雾弥漫开来时，南加州的圣盖博山脉很美。《雾色山脉》是一个很好的使用波浪图案的例子。

The San Gabriel Mountains in Southern California are beautiful when the fog rolls in. This piece, "Misty Mountains", is a perfect example of utilizing the wave pattern.

图案 Patterns

树木，分形图案
Trees, Fractal Pattern

树木的枝干和根系是自然中分形的例子。《飞》是一个在压花艺术中使用分形图形的实例。

Tree branches and roots are examples of fractal patterns in nature. "Flying" is one sample of how to apply fractal pattern into pressed flower art.

镶嵌图案
Tessellation Pattern

镶嵌是通过在整个平面上重复铺贴块状而形成的图案。马赛克和拼布是两种很常见的运用方法。

Tessellations are patterns formed by repeating tiles all over a flat surface. Mosaic and quilt are two typical applications.

我们可以设计简单的元素，然后像拼布一样将它们组合在一起。这种设计是类似/对比的完美示例。像其他任何设计一样，我们仍然需要有主题，这样作品才有意义，而不仅仅是美观。

We can design simple elements and then assemble them together like a quilt. This type of design is a perfect example of similarity/contrast. Just like any other designs, we still need a main focus so the picture has some meaning instead of simply aesthetics.

丰盛 Abundance

图案 Patterns

花窗玻璃式设计
Design Stained Glass Style

彩色花窗玻璃艺术总是令我惊叹不已。我非常喜欢观看教堂的窗户，甚至只是呆看我的彩色玻璃灯罩。花窗玻璃也是另一种使用图案的方法。多年前我看过一个专业人士用彩色玻璃技术制作玻璃压花的挂饰。它涉及到包裹金属和焊接。虽然在大学时代，我就已经会焊接，但我觉得整个过程都很复杂。完成的挂饰非常漂亮，但也费工和需要特殊技术。

2011年我前往英国牛津参加压花春季研讨会。在那里我学会了一种制作彩色玻璃花窗式压花卡片的方法。这堂课给了我很多启发，让我进一步探索更简易轻松的花窗式压花设计。

Stained glass art has always amazed me. I just love to look through windows at church or even just looking at my stained glass lamp shade. Stained glass is another way applying pattern in designs. Many years ago, I have watched a professional making glass pressed flower hanging objects with stained glass techniques. It involved with foiling and soldering. Although I have used a soldering iron working with electronics when studying in engineering school, I thought the whole process is very complicated. The finished projects are very pretty but also labor intensive.

I traveled to the UK in 2011 to attend the Pressed Flower Guild spring conference in Oxford. There I learned a way to make stained glass pressed flower card design. The class gave me ideas that allowed me to explore easy pressed flower stained glass projects further.

运用纸胶带制作简单彩色花窗玻璃设计
Easy Stained Glass with Washi Tapes

材料 Materials

1. 金属色或黑色细和纸胶带（3mm宽度）
2. 冷裱膜
3. 各种薄压花
4. 相框（7寸或8寸）
5. 热胶枪

1. Skinny Washi tape in metallic or black color (1/8" or 3mm)
2. Self-adhesive lamination sheet
3. Thin assorted pressed flowers
4. Picture frame 5"×7" or 8"×10"
5. Hot glue gun

制作方法 Procedure

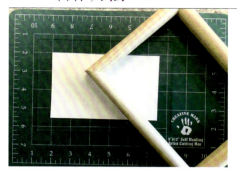

将玻璃与框架分开。清洁玻璃。用钳子将所有卡子从框架中拉出。

剪一片冷裱膜。7英寸框大约9cm×14cm。8寸框大约15cm×20cm。

Separate the glass from the frame. Clean the glass. Pull all the points out from the frame with a pair of pliers.

Cut a piece of self-adhesive film. For 5"×7" frames, cut to 3.5"×5.5"×5.5". For 8"×10"frames, cut to 6"×8".

1

2

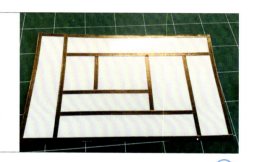

从四周开始，先贴四周。外圈不要有白边。

Start from all the sides, glue the washi tape around. Do not leave any white edges on the sides.

继续随机粘贴一些方或长方形。您必须测量花朵以确定形状的大小。7英寸框需要使用较小的花朵。可以放在带标记的垫板上操作确保把纸胶带贴直线。您可以提起纸胶带以校正位置。完成所有形状后，用手指按实纸胶带。不要用力摩擦，以免损坏纸胶带。把压花摆放在形状的框架中做设计。用手机拍照。

Continue to glue some boxes at random. You have to measure your flowers to determine the box sizes. Try smaller flowers for smaller frames. Make sure the lines are straight. You can lift the washi tapes up to make corrections on positions. Once you have all the boxes completed done, press down all the washi tapes with your fingers. Do not rub too hard and to damage the tape. Place flowers into the boxes for your design. Take a picture for reference with your cellphone.

3

图案 Patterns

4 把背面的离型纸撕开。把冷裱膜粘的一边朝上放。

Peel off the protection paper in the back. Place the lamination film face down so the sticky side faces up.

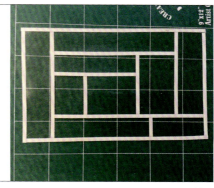

5 按照之前手机拍的照片把压花脸朝下放在形状中。

Place the flowers face down into all the boxes according to the picture you have taken before.

6 在一张白纸上画上玻璃的形状。对于7英寸镜框，在所有边上留出2cm的空间，并在玻璃形状的中间绘制9cm×4cm形状。对于8英寸镜框，在所有边上留出2.5cm的空间。

Draw a shape of the glass on a piece of white paper. For 5"×7" frames, leave 3/4" space on all sides and draw the 3.5"×5.5"×5.5" shape in the middle of the 5"×7". For 8"×10" frames, leave 1" space on all sides.

7 将贴好花朵朝下的冷裱膜放在您绘制的中间长方形状上。然后根据图纸放置干净的玻璃。这将确保您的线条对得整齐。

Place the self-laminating film with flowers face down onto the middle rectangular shape that you have drawn. Then, place the clean glass down according to the drawing. This will ensure your alignment is correct.

8

把玻璃翻过来，用手指从中间往四边按压冷裱膜。把空气排出。可以用缝纫的针刺穿小孔把气泡中的气排出来。

Flip the glass over and press down the self-lamination film from the center to all 4 sides and corners. Squeeze out air bubbles as you go. Use a sewing needle to poke tiny holes to let trapped air out if needed.

9

用热胶枪在镜框背后的四个角挤少量胶。把玻璃脸朝下贴到镜框上。

Put a small dot of hot glue on each back corner of the frame. Place the glass face down to glue onto the frame.

10

在四角上再各加一点胶确保玻璃贴牢。

Add a drop on the corners again to ensure the glass is secured.

彩色花窗玻璃玫瑰
Stained Glass Rose

材料 Materials

1. 金属色或黑色细和纸胶带（3mm宽度）
2. 高透光PVC薄片（10cm方形）
3. 深浅两种不同的玫瑰花瓣
4. 白色卡纸或水彩纸
5. 冷裱膜（11cm方形）

1. Skinny Washi tape in metallic or black color (1/8" or 3mm width)
2. Dura-lar clear or clear craft plastic sheet (4" or 10 cm square)
3. Rose petals (two shades of pink or other colors)
4. White card stock of water color paper
5. Self-lamination film (4.25" or 11 cm square)

制作方法 Procedure

1 用纸胶带在透明PVC薄片上转圈形成玫瑰的形状。

Use washi tape to circulate the 4" clear sheet forming pattern for the rose.

2 用深浅不同的玫瑰花瓣交错地贴在背面的空白处（注意，花瓣朝下，这样我们从另一侧看到的是正面）。

Alternate dark and lighter rose petals in different areas in the back side. Note, glue the petals face down so we see the front of the petal the other side.

3 贴上冷裱膜，边缘留3mm。

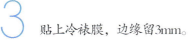

Apply self-lamination film on the top. Leave 1/8" or 3mm on the edges.

4 翻转。贴在背景纸上。在玫瑰边缘贴一圈纸胶带。

Flip over. Glue onto the background paper. Apply washi tape on the edge of the rose to finish.

彩色花窗玻璃模板
Stained Glass Pattern

　　花窗玻璃有很多不同的设计。可以画出自己感兴趣的东西，或者可以使用免费的图案。当然也可以购买花窗玻璃的图案。

　　非常感谢Chantal Pare给我许可使用她免费的彩色玻璃图案（http://free-stainedglasspatterns.com/）。

There are many different designs. One can draw outlines on something we are interested in or can use the free patterns found online. One can certainly buy stained glass patterns also.

Many thanks go to Chantal Pare for giving me permission to use her free stained glass patterns. (http://free-stainedglasspatterns.com/).

下面这个模板是我画的，大家可以自由使用。

The one below is one I have drawn. We can use it freely.

图案 Patterns

材料 Materials

1. 两份图像（一份正面，一份反面）
2. 双面贴
3. 背景用的浅色花瓣或叶子
4. 几种不同深浅的紫色花瓣大片的绿色叶子

1. Two images (one right side, one reserved)
2. Double sided adhesive
3. Light color petals or leaves for background
4. Several different shades of purple petals

    Large mass green leaves

按照模板作画 Making Pressed Flower Picture According to Pattern

1 将图案描到一纸上。反面也描一张。也可以将图案扫描到计算机上打印一张图像；然后翻转图像并再次打印。将双面贴粘在图像的正面。这将会是我们制作压花画的底纸。将双面贴粘在反转图像的背面。

Trace the pattern onto a piece of paper. Make a reverse trace as well. One can scan the pattern onto computer print one image. Then use computer software too; flip the image and print again. Place double sided adhesive on the front of the right side image. Place double sided adhesive on back of the reversed image.

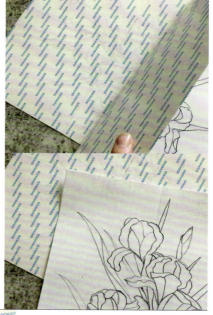

2 把图形从反转图上一片片地剪下来，把花瓣或叶片根据图案贴上。

Cut out shapes from the reversed image. Glue petals or leaves according to the picture.

3

根据形状把多余的花瓣或叶片修剪一下。把形状贴到正面图案上。

Trim the extra off according to the shape.Glue the shape onto the right side of the image.

4

完成粘合花瓣和叶子。

Finished gluing the petals and leaves.

制作花窗玻璃式的凸面线条
Making Stained Glass Liked Outlines

有几种方法可以制作轮廓。

制做轮廓的一种方法是使用三维凸面涂料。在压花图上绘制轮廓之前，先在草稿纸上练习。绘制完成后至少搁置24小时以确保干燥。

There are several ways to make the outlines.

One way to make the outline is to use dimensional paint. Practice on a scratch paper first before painting the outline on the pressed flower picture. Set it aside for at least 24 hours before touching to ensure the outline paint is dried.

另一种方法是使用胶水笔绘制轮廓，然后在上面倒入微细的手工用黑色粉末（类似凸粉）。多余的粉末可以倒在一张纸上，然后倒回瓶内。你可以一小片地方一小片地方地来，直到完成整个画面。

Another way is to use a glue pen to draw outlines and then pour micro fine black powder on the top. Catch the extra powder with a piece of paper. You can work on one small area at a time.

设计自己的图案
Design Own Pattern

我们可以为自己的作品设计属于自己的图案。

We can design our own patterns for our picture.

04

多媒介
Multimedia

压花适合混入各种媒介。我已经在《压花艺术（中级）》书中讨论过使用各种不同媒介的背景作为媒介。在本章里，我们将看到将压花进一步混合到不同的媒介之中从而制作出艺术精品。

Pressed flowers works very well with all kinds of medium. I have discussed about using different mediums for backgrounds in Book 2. Here, we will see that we can further use different materials mixed into pressed flower art.

水晶花瓶插花
Arrangement in Crystal Vase

材料 Materials

1. 水彩纸 (20.5cm×25.5cm)
2. 硫酸纸
3. 铅笔（HB）
4. 白色铅笔或白炭笔
5. 压花角堇、梗、叶

1. Watercolor paper (8"x10")
2. Translucent vellum paper
3. Pencil (HB)
4. White charcoal or color pencil
5. Pressed viola, stems, leaves

多媒介 Multimedia

制作水晶花瓶 Make Crystal Vase

1. 把硫酸纸覆盖在图片上,用铅笔把图连同阴影都描出来。

Use vellum paper over the image to trace the image with pencil including the shading.

2. 用铅笔画一个椭圆水线。用白色炭笔或彩铅笔打出花瓶的亮点。

Use pencil to draw an oval waterline. Use either white charcoal pencil or white color pencil to paint a highlight spot.

制作插花 Make Arrangement

1. 先在硫酸纸上描好水面这个椭圆,然后把椭圆剪下来。在水面上粘一些梗。注意,梗随意地布置在水面,不要排队般。梗的上端伸向瓶口。把椭圆形水面形状贴好。注意我们这是从瓶子的背面贴。

Trace the waterline oval to a piece of vellum and then cut it out. Arrange a few stems on the waterline. Notice that the ends are on different positions randomly – not lining up. The top of the stem needs to point out of the vase. Cover the waterline with the cut out piece. Note: We are gluing on the back of the vase.

多媒介　Multimedia

制作水晶花瓶 Make Crystal Vase

1

把硫酸纸覆盖在图片上,用铅笔把图连同阴影都描出来。

Use vellum paper over the image to trace the image with pencil including the shading.

2

用铅笔画一个椭圆水线。用白色炭笔或彩铅笔打出花瓶的亮点。

Use pencil to draw an oval waterline. Use either white charcoal pencil or white color pencil to paint a highlight spot.

制作插花 Make Arrangement

1

先在硫酸纸上描好水面这个椭圆,然后把椭圆剪下来。在水面上粘一些梗。注意,梗随意地布置在水面,不要排队般。梗的上端伸向瓶口。把椭圆形水面形状贴好。注意我们这是从瓶子的背面贴。

Trace the waterline oval to a piece of vellum and then cut it out. Arrange a few stems on the waterline. Notice that the ends are on different positions randomly – not lining up. The top of the stem needs to point out of the vase. Cover the waterline with the cut out piece. Note: We are gluing on the back of the vase.

多媒介　Multimedia

2 让我们看一下右面这幅照片。可以看到折射的影像。吸管在水中好像和在水上折成两段。所以我们水面和水下的枝是分开来贴的。

Take a look of this photo on the right. We can see the effects of refraction. It looks as if the straw is broken. Therefore, we are creating the same effect by arranging the stems above the waterline and below waterline separately.

3 贴水下的枝。每枝水面上的枝都必须有水下的。贴得稍微和上面错开一点，但不要离得太远。

Arrange the stems below waterline as shown. For every stem above the waterline, there must be a stem below. Arrange it slightly off but not too far away.

4 把花瓶翻过来。这边就是我们要用的。需要修改调整的地方现在调整好。用钝铅笔把底纸上放花瓶的地方画上阴影。我通常下面留4~5cm的空隙。把花瓶放好，用一点胶固定起来。

Turn the vase over. This is the side we use. Make adjustments if necessary. Use the broad side of pencil to draw some shades on the background where the vase will stand. I usually leave about 1.5" to 2" space from the bottom. Place the vase on the shade and use a little bit of glue to secure.

5

把三色堇和角堇贴好。写好日期、签名。完成。

Arrange pansies and violas. Date and sign to finish.

蓝白陶瓷花瓶插花
Arrangement in Ceramic Vase

材料 Materials

1.水彩纸 (20.5cm × 25.5cm)

2.薄和纸或薄宣纸

3.天蓝色水彩或染料

4.双面贴（最少6cm宽）

5.小的有色彩的薄花，如绣球

6.压花迷你玫瑰，需要多色，花、花蕾和叶子，或其他各色连枝花

7.压花蕾丝或满天星

8.选择：珍珠白色细凸粉和热风枪、微型透明玻璃珠，或者白色极细闪粉。水彩纸 (20.5cm × 25.5cm)

1.Watercolor paper (8"×10")

2.Tenchgucho paper

3.Sky blue dye or watercolor

4.Double sided adhesive film (minimum 2.5" wide)

5.Small colorful and flat pressed flowers such as hydrangea

6.Pressed mini roses of mixed colors (need several colors, leaves or we can use several different flower stems)

7.Pressed Queen Anne's lace (or baby's breath)

8.Choose either pearl white fine embossing powder with heating tool, clear micro glass beads or white micro glitter

制作陶瓷瓶 Make Ceramic Vase

1 剪一块比花瓶稍长的63mm宽双面胶带。将白色凸粉倒在双面胶带上。然后使用热枪来熔化凸粉。不要过热，否则会产生气泡。趁热再沾凸粉，尤其是中间地带需要更多的凸粉。使用热风枪再次熔化凸粉。如果没有热风枪，可以尝试将微细白闪粉末撒到双面胶带上，然后用手指细细揉，使闪粉固定好。或者不用闪粉，可以尝试微细玻璃珠。

Cut a piece of 2.5" wide double sided tape slightly longer than the vase in section. Pour white embossing powder all over on one side of the double sided tape. Use a heat gun to melt the embossing powder. Do not over heat or you will see bubbles. While still hot, try to get more embossing powder on especially in the center. Use the heat gun to melt the embossing powder again. If you do not have a heat gun, you can try rubbing micro fine white glitter powder onto the double sided tape. Alternatively, we can try micro glass beads.

2 剪一块比花瓶稍长的63mm宽双面胶带，贴在白色卡纸上。贴几朵蓝色绣球（或你喜欢的别的彩色薄花）。

Place a double sided adhesive slightly larger than the vase on a piece of white card stock. Arrange some blue hydrangea (or any flat colorful flowers you want to use).

3 把有凸粉的双面胶保护纸撕下，贴在绣球上面。

Peel off the protection paper from the embossed one and glue it on top of the hydrangea.

4 使用花瓶图片来把形状剪出来。

Use the vase picture as template to cut the shape out.

107

制作背景 Making Background

使用染料来染典具纸是我很喜欢使用的方法。参看《压花艺术（中级）》第关于使用染料制作背景。还有一种很简单的方法也可以获得彩色典具纸的效果，就是在典具纸后面垫一张彩色纸。

Use dye to prepare tengucho paper is one way I like to use. See Book 2: dye background method. One can also simply use a color paper behind the tengucho paper to achieve colored background.

当纸完全干燥之后，揉成一团，再小心地打开展平。这是为了把僵硬的纸张纤维打软以防止皱褶产生。

After paper is completely dried, squeeze it into a ball and then extend it out carefully. This is to break the stiffness of the paper fiber to prevent wrinkles later on.

如果打算真空密封完成的作品，裁一块白色棉垫放在我们刚刚着色的薄纸背面。如果不打算真空密封，请使用一张水彩纸作为衬底。使用软粉彩对背衬材料稍微着色，为花瓶坐的地方打一些阴影。

If we are vacuum sealing the finished project, have a piece of white low pile cotton batting as back of the thin paper we have just colored. If we are not vacuum sealing, use a piece of watercolor paper as the back. Use soft pastel to color the backing material to create some shading for the picture where the vase will stand.

多媒介　Multimedia

制作插花 Making Arrangement

1

用少量的胶把花瓶固定好。然后在花瓶口贴好叶子。

Use a little glue to secure the vase in position. Then arrange some mini rose leaves around the top of the vase.

2

粘2~3个迷你玫瑰花蕾来定瓶花的高度。请注意，花瓶之下的空间和瓶花之上的空间应该大致相等。

Glue 2-3 mini rose buds to define the height of the arrangement. Notice that the space below the vase and above the arrangement are supposed to be about equal.

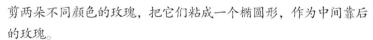

3

剪两朵不同颜色的玫瑰，把它们粘成一个椭圆形，作为中间靠后的玫瑰。

Cut two different colors of roses in half and arrange them to form an oval shape for mid ground back roses.

4

贴两朵不同颜色（最好是不同色于后面的）重叠于前面。

Glue two different colors (best to use different from the two back roses.) overlapping the mid ground back ones.

109

5 粘4~5个不同颜色的玫瑰在中心前方重叠。剪2~3朵玫瑰插在两侧后面,使瓶花更加丰富。然后将蕾丝花插入。在"桌子"上添加几枝玫瑰花和蕾丝花。乡村风格瓶花就完成了。在薄纸上签名是一个挑战。我们需要在背面衬一张普通纸和垫板。使用无酸凝胶滚珠签名笔。

Glue 4-5 different colors of roses overlapping in the center front. Cut 2 -3 roses to place in on the sides making the arrangement fuller. Then insert Queen Anne's lace all over. Add a couple stems of roses and Queen Anne's laces on the "table". A country style arrangement is finished. Signing on thin paper is a challenge. We need to have a piece of regular paper on the back and on hard surface. Use acid free gel rolling pen to sign.

多媒介　Multimedia

雨过天晴花瓶插花
Arrangement in Raindrop Vase

材料 Materials

1. 水彩纸 (20.5cm × 25.5cm)
2. Distress Oxide颜料
3. 喷水壶
4. 手工垫、烘培垫或任何防水垫
5. 混色工具
6. 压连枝带叶雏菊

1. Watercolor paper (8"×10")
2. Distress Oxide
3. Water spray bottle
4. Craft sheet, silicone baking sheet or other waterproof sheet
5. Blending tool
6. Pressed daisies with stems and leaves

制作雨过天晴花瓶 Making the Raindrop Vase

先从103页把花瓶描好。，然后剪下形状。把花器的外形描在一张水彩纸上。

Cut the vase pattern trace from page 103 out and trace the shape of it on a piece of watercolor paper.

1

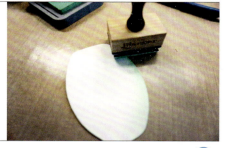

2

使用一个混色工具，沾一点颜色（cracked pistachio），然后从边缘开始以转圈的方法涂往中间。

With the blending tool dip on the color pad (cracked pistachio), apply the colors onto the vase shape in circular motion from the edge toward the center.

可以在左边涂上不同的颜色（broken china）。要记得用另一个干净的海绵垫沾不同的颜色。

3

We can apply a different color (broken china) on the left using the same method. Make sure to use a different clean foam pad when change color.

111

4 用喷壶喷一些水在涂好颜色的花器上。不要喷太多。

Use a spray bottle to spray some water onto the colored vase shape. Do not saturate.

5 等待干燥。可以使用一把热风枪加速干燥时间。

Wait for it to dry. We can use a heat gun to speed up the drying process.

制作背景 Making Background

1 在硅胶垫上抹一些颜色（faded jeans, iced spruce, worn lipstick）。

Apply some colors (faded jeans, iced spruce, worn lipstick) on the silicone mat.

2 喷一些水。可以看到颜色遇水起了反应。

Spray some water. We can see the colors reacting to water.

多媒介　Multimedia

3

把背景的水彩纸覆盖在颜色上，来蘸色。

Dip the background watercolor paper onto the color pools.

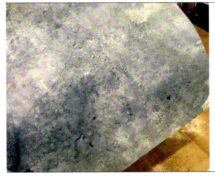

把纸拿起来看看。没有什么对与错，每张都是独一无二的。用热风枪或吹风机快速吹干。然后喷一些水，让颜色起氧化效果。再次吹干。

4

Lift the paper up to see how it looks. There is no wrong way. Each paper is unique. Use heat gun or hair drier for fast drying. Spray some water after dried for oxide look. Dry it again.

5

如果需要，在硅胶垫干净和干燥的地方加点颜色，然后喷水。把背景白色部位拿去蘸颜色。然后再次吹干、喷水、吹干。把少量地方蘸粉红色，吹干、喷水、吹干。这样的背景看起来会很有层次感。重复，直到背景色彩满意为止。用混色工具涂一些灰色或蓝灰色在花瓶所要站立的部位打一打阴影。

Add more color to a clean and dry area of the silicone mat if needed and spray water. Dip the white area onto color. Dry and spray water for oxide effect as before. Dip areas onto the pink color to add some variation of color. Dry and spray water for oxide effect as before. Work on the background to your satisfaction. Use a blending tool to add some shading to the area where vase will sit.

压连枝晶菊 Pressing the Daisies With Stems

1 瓶花用的花需要有梗。白晶菊连枝一起压会比单花正面压的看起来更自然。

Pressed flowers used for vase should have stems. Daises pressed with stems together would look more natural than full faced flowers.

2 采集第一天开的白晶菊。第一天开的,花心比开过几天的薄很多。用手指按压它的一边强制花枝往另一边倒。将花脸朝下。白晶菊背面用的压花纸张需要非常柔软。面巾纸是不错的选择。盖一层海绵。然后用还原好的干燥剂板夹住。用约5kg的压力。花应在1~2天内干燥。

Collect daisies opened on the fist day. Centers are much thinner than the ones opened for several days. Use our finger to force the stem on one side by pressing it. Place the flower face down. The paper used for the back of the daisy must be very soft. Paper towel or facial tissue are good choices. Top with a layer of foam. Sandwich with active desiccant boards and apply about eight to ten pounds of pressure. Flowers should be dried in one to two days.

制作插花 Making Arrangement

1 当我们观察真实的瓶花,就会在花间的空隙处看到一些花的背面。所以我们在制作的时候也要这样安排。

You would see some flowers' backside in a real arrangement. Hence, we arrange some in that orientation.

2 把花瓶安放好。确保所有的枝都在花瓶"里面"。再添加少许花使花瓶更加丰富。

Place vase in position. Make sure all the stems are "inside" of the case. Add a little more flowers to make the arrangement fuller to finish.

多媒介 Multimedia

05

静物
Still-life

　　静物是一件由不动物体组成的艺术品。静物捕捉的是生命瞬间的本质。从一开始，艺术理论家就嘲笑静物画，为"低级"流派艺术家在没有运用想象力的情况下复制自然。确实，静物画常常紧密地模仿现实。但是它们经常也包含许多更深层的含义。宗教、经济、科学、政治信仰和联想都可以嵌入到花朵或水果的图像中。

　　一个好的压花静物设计不仅要考虑物体本身，还要考虑背景和光照条件。

A still life is a work of art which represents a subject composed of inanimate objects. Still life is the essence of life captured at the moment. From its inception, art theorists derided still life as a "lowly" genre artists merely copied nature without using their imagination. Indeed, still life paintings often closely imitate reality. But they frequently also contain many deeper meanings. Religious, economic, scientific, political beliefs and associations may all be embedded in a single image of flowers or fruit.

A good still life design in pressed flower will not only consider the objects themselves but also the background and lighting conditions.

花篮
Flower Basket

  篮子是由交织的材料制成。通常使用木头、竹子、麦杆、其他草、草皮、树枝、柳条或藤条来做篮子。要设计一个逼真的篮子，我们必须展示编织图案和其材料的样子。即使是再逼真的篮子，单独一个篮子的静物画也永远不会令人兴奋，因为静物篮子的生命是由篮子里的生活材料组成的。

A basket is made of interwoven material. Wood, bamboo, wheat, other grasses, rushes, twigs, osiers, or wicker are often used to make baskets. To design a realistic basket, we have to show the woven pattern and suggest of the material. A basket alone is never that exciting even the best detailed basket. A basket is made alive by the life materials it contains.

  下面这些是制作花篮时要考虑的事项：
  花篮是用于什么场合？这种场合有时需要特殊的颜色或花朵类型，可能还需要特殊风格的篮子和布置。然后选择随场合搭配的主花。选择取决于花的语言、季节或传统。花的大小取决于作品的大小。选择与主花形成某种颜色对比或互补的配花。与鲜插花不同，单色创作对压花来说并不讨好。布置风格取决于场合和环境。例如，正式聚会中使用的插花应更为正式，而花园派对中使用的插花应更为随意。选择与布置样式统一的花篮。使用真实的插花图片作为压花静物画的参考是一个好方法。

These are the things to consider when creating a flower basket.

What is the occasion? The occasion sometimes demands for particular color or type of flowers. The occasion might also demand for particular style of basket and arrangement. Select the main flower that will go with the occasion. Selection depends on language of flower, season, or tradition. Size of the flower depends on the picture size. Select companion flowers that provides some color contrast or compliment to the main flower. Unlike fresh arrangements, mono color theme does not work will pressed. Style of arrangement depends on the occasion and surroundings. For example, arrangements used in formal party should be more formal but arrangements used in a garden party should be more casual. Select the basket that will go with the arrangement style. It is a good idea to use real flower arrangement pictures as reference for pressed flower still life creation.

　　关于花的一些事情。
　　要展现真实的插花要求以许多角度压花。至少需要正面和侧面。有45°角压花更好。最明显的是雏菊。如果在压连茎的，按压并尝试以一定角度弯曲梗子，压出来则很可能会有一定角度。主花的质量非常重要。如果压花快用完了，最好等待新的花压干，而不要使用一些次等花。

A few things about the flowers.

A realistic flower arrangement demands flowers pressed in many angles. At minimum, the full faced and side views are needed. It is desirable to have some pressed at 45-degree angle. The most visible ones are daisy type of flowers. If we press them with stems on and try to bend them at an angle, they will most likely be pressed out with an angle. The main flower quality is extremely important. If we are running out of flowers, it is better to wait for new flowers to dry than use some second rated flowers.

春之庭院花篮 Spring Garden Basket

我将详细介绍篮子的结构。 我相信现在每个人都可以把花材装满篮子。 这张画的背景是一块绢，为了不分散观众对篮子里花材的注意力，我把大部分花园中的薰衣草安排在绢的下边。

I will go over the basket construction in detail. I am sure everyone can fill the basket by now. This picture's background is a piece of silk. I have arranged some lavenders under the silk so they will not distract the flowers inside the basket.

材料 Materials

1. 28cm×35.5cm素描纸或打印纸
2. 28cm×35.5cm绢
3. 浅绿色、草绿色、深橄榄色云龙纸
4. 双面胶
5. 花材：

1. 11"×14" sketch paper
2. 11"×14" silk
3. Light green, grass green, dark olive mulberry paper
4. Double sided adhesive
5. Pressed materials:
   a. Gerbera daisy (light pink, purple)

a.非洲菊（浅粉色，紫色）
b.薰衣草
c.蕨芽
d.小红玫瑰
e.白鳞托菊
f.Hardenbergia
g.勿忘我

b.Lavender
c.Fern buds
d.Small red rose
e.White daisy
f.Hardenbergia
g.Forget-me-not

注意：有许多适合制作篮子的花瓣和叶子。我最喜欢的是浅色或褪色的非洲菊花瓣。花瓣尺寸相对均匀，因此易于使用。

Note: There are many flower petals and leaves suitable for making baskets. My favorite is light color or faded gerbera daisy petals. The petal sizes are relatively even making them easy to work with.

1

我画了一个篮子的形状。篮子的大小取决于画面的大小。相对于我的28cm×35.5cm画面，我的花篮尺寸高度为18cm。

I have drawn a basket shape. The size of the basket depends on the picture size. For my 11"×14" picture, my basket size is 7" tall.

2

我把花篮的图片在电脑上翻转打印。
把提手和篮子分开剪。不要剪得太靠近黑线，四周留下一小点边。

I flip the basket horizontally to have a reverse image on my computer and print out.

Cut the handle and the basket separately but leave small spaces all around the edges.

121

3

把双面贴粘在白色的那边。篮子和提手都贴好。

Apply double sided adhesive to the white side of the basket and handle all over.

4

一次提起并折叠一小段离型纸（足以进行粘贴）。从左边开始，竖着贴非洲菊花瓣。贴完一竖排再贴下面一竖排。排列要相对直。

Lift and fold the double sided adhesive protection paper a small section at a time (just enough to work on). Glue the gerbera daisy petals from left one column at a time. Make the columns relatively straight.

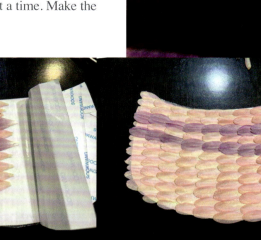

5

3条线是从一侧到另一侧的把手。握持部分被包裹。因此，我们需要以不同的方式粘贴这些部件，才能看起来像篮子的结构。

The 3 lines are the actual handle that goes from one side to the other. The grip part is wrapped. Therefore we need to glue these parts differently to resemble the basket construction.

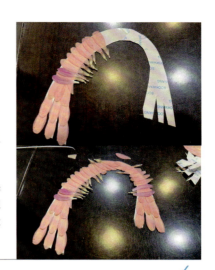

6

反过来按照图纸剪齐篮子和把手的边缘。

Flip and trim the basket and handle according to the drawing.

用花材填充篮子时，因为我们没有制作篮筐的内部和边缘，请确保中心部分完全填充。如果我们的画需要一个相对较空的篮子，请将这些部分分别做好。请记住，由于光线条件，内部应显得较暗。

When filling the basket, make sure to fill the center part completely because we did not make the inner portion and the rim of the basket. If we need to present a relatively empty basket, make sure to make those parts. Remember that inside should look a shade darker because of the lighting condition.

其他花篮 Other Flower Baskets

市面上有很多不同形状、大小的花篮。这里我又画了一个。

There are so many baskets in different shapes and sizes. I have drawn another one that I like to use.

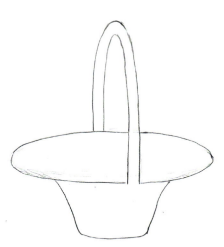

我用勋章菊的叶子做这个篮子。由于叶子比我之前展示的小非洲菊大很多，因此我们可以通过改变叶子的正反两面来做一些有趣的纹理。

I used gazania leaves to make this basket. Since the leaf size is a lot bigger than the small gerbera daisy I have shown before, we can simply the basket by alternating the front and back of the leaf to make some interesting patters.

很多材料都可以拿来制作花篮。这是一个蘑菇花篮的范例。

Many materials can be used to make baskets. Here is an example of mushroom basket.

有时我们并不需要制作非常真实的花篮，我们可以制作出看起来像花篮的篮子。

Sometimes we don't need to have such realistic baskets. We can also make some that look like baskets.

上面这个花篮的材料是一片虫咬的叶子。

I used a bug bitten leaf for the basket above.

　　黄蜂巢由植物材料组成。它具有非常有趣的线条图案。深秋或冬天气温达到冰冻温度后，我们可以收获黄蜂巢。

Wasp nest are comprised of plant materials. It has very interesting lines of patterns. We can harvest the nest in late fall or early winter after freezing temperature is reached.

　　有时我们不想制作非常真实的篮子。我们对图案和形状更感兴趣。在这里，我用一个个小的铁树芽制作篮子。它的手柄由铁树芽和中生代铁树叶组成。

Sometimes we do not want to make very realistic baskets. We are interested with patterns and shapes. Here, we make a basket with very young sago palm leaves. Its handle is made of older sago palm leaves.

静物 Still-life

花瓶
Flower Vases

我们之前做了几个花瓶。 在这里，我介绍了一些简单但有效的制作花瓶的方法。 并非每幅压花艺术画都需要非常真实。 我们可以有一个简单的一支花插入瓶中，也可以有复杂的插法。 与鲜花插花不同，我们不需要相同的花梗或叶子，可以混合搭配。 但是，所有事物都必须看起来生动有趣（例如，作品必须遵循物理定律，让它看起来不会翻倒）。

We have made a few flower vases before. Here I am presenting some simple but effective ways making flower vases. Not every pressed flower art picture need to be realism. We can have simple one stem arrangement as well as complicated arrangements. Unlike fresh flower arrangements, we do not need to have the same flower stems or leaves. We can mix and match. However, everything must look alive and believable (for example, arrangements must follow law of physics so it does not look tipping over).

鳄梨果壳看起来和陶器差不多。做一个古朴的陶瓶，插一支花。

The avocado shell resembles to pottery. Make a simple earthenware vase and arrange a single stem flower.

127

用一块浅绿色的硫酸纸剪出花瓶形状，我在纸背面用油粉彩画棒涂了几行直的线。然后再为小花瓶插花。

Cut a vase shape with a piece of light green vellum. I painted a few lines running vertically with white oil pastel on the back of the vellum. Arrange flowers for the small vase.

把绣球花瓣剪碎再贴在花瓶形状上使花瓶看起来像冰裂纹瓷器。

Cut up the hydrangea petals to glue on the shape of a vase resembles the cracked porcelain.

静物 Still-life

结合绿叶，苎麻和红脉酸模叶子制成一个有趣的花瓶。

Combining green leaf, ramie, and bloody dock leaves to make an interesting vase.

用3片褪色大型玫瑰花瓣组合成有旋转花纹的陶瓶。

Use three faded old large rose petals to make a vase with painted spiral lines.

使用几根茎秆来定义一个高脚香槟杯。我们可以做出抽象的花艺。

Using a few stems to define a tall champagne glass. We can make an abstract arrangement.

我们甚至可以画背景和花器来制作混合媒介压花艺术。

We can even paint the background and the vase to make mixed media pressed flower art.

花盆
Flowerpots

当我们在室内摆放盆栽植物时,我们经常将盆栽植物与盆一起放置在一个装饰用的篮子或盆里面。这种装饰用的篮子通常有塑料衬里,以防止水流出来损坏家具。如果装饰的容器是花盆,它盆底不会有洞让水流出。如果容器材质本身不具备防水能力,中间会有塑胶隔层。

When we display potted flowers inside, we often use a decorated basket or pot for the outside and place the potted plants with pot together inside. The baskets usually have plastic linings to prevent water running out damaging our furniture. If the decorative container is a pot, it does not have the hole to let water out. If the materials is not waterproof, plastic lining is applied.

制作这些木制花盆非常容易,只需剪一块树皮即可。 但是,作品的注意力要集中在植物的活泼性上。 我们不仅需要压许多不同的角度和不同开花阶段的花。 我们还需要很多叶子。

Making these wooden pots are very easy, just cut a piece of bark. However, the attention is focused to the liveliness of the plant. Make sure to press flowers with many different angles, and different stages of bloom. We also need lots of leaves.

压仙客来花时,将其后面的梗剃去一些。 用微波炉加热30

秒钟，然后取出放掉蒸汽。 根据干燥情况调整时间。 一直用微波炉加热，直到干燥程度达到80%~90%。 将花朵放在干燥板中以使其完全干燥。关于花蕾，可以稍微修剪一下背面，也可以用微波炉压。 对于叶子，它们在微波炉中的效果不佳，绿叶会因加热而变得有点橄榄色。 我们取一小块粗糙的砂纸（我用60粒粗砂纸），用手指将叶背面按在沙纸上，在后背上造成许多细小的孔，然后在干燥板中压叶子。

To press cyclamen, shave the back of the stem off. Microwave 30 seconds and let the steam off. Adjust time according to the drying condition. Microwave all the way until 80%-90% dry. Place the flowers in desiccant press for complete dry. For buds, we can trim the back side off a little and press with microwave also. For leaves, they do not do as well in the microwave. Greens will turn somewhat olive with heat. We get a small piece of coarse sandpaper (60 grits is what I use). Press the leave back against sandpaper with your finger to cause lots of tiny punctures on the back. Press the leaves in desiccant press.

苔藓盆将室外的感觉带入室内。 有多种形式的苔藓盆，干燥染色青苔覆盖在有塑料衬里的盆在许多地方都可以买到。我们的苔藓盆是把苔藓拉直一行行地贴在纸模上。

Moss pot brings outdoor to indoor. There are many forms of moss pots available. The dried and dyed moss covering a pot with plastic lining is available in many places. Glue moss on a pot shape. It is earlier to straighten moss in lines to glue.

静物 Still-life

具有光影效果的花盆可以很容易地用在同一片叶子上具有自然两种颜色的叶子制成。

Pots with light and shadow effect can be made easily with leaves that has natural two colors on the same leaf.

由于树叶全年受阳光照射的方式，许多树叶在秋季都具有此属性。 在秋天收集较大颜色较浅和较深在同一叶子。

Many leaves have this property in the fall due to the way the leaves receive sun throughout the year. Collect larger leaves with both lighter and darker colors in the fall.

06

幻想类
Fantasies

花仙子
Fairies

幻想类 Fantasies

花仙子 Fairies

从小我就喜欢各种童话，仙子的魔法使我着迷。

Since I was child, I have loved all kinds of fairies since I was a child.

我喜欢用人类的比例画仙子。我们使用头的大小作为比例，成年女性的身高通常为7.5个头大小。使用人体模型作为绘图模型会有所帮助。

I like to draw the fairies using natural proportions of humans. We use head size as the scale. Adult female height is usually 7.5 head size. It helps to use a mannequin as the model.

我首先根据人物的姿势和比例绘制一个简笔画，然后填上"肌肉"。接下来绘制服装、头发和珠宝（如果有），最后画上翅膀。

I first draw a stick figure according to the position and scale. Then, I fill the figure with flesh. Next, I draw the cloth, hair, and jewelry. Finally, I draw the wings.

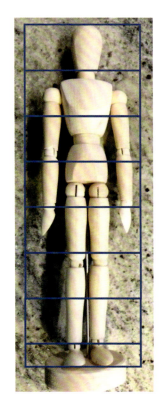

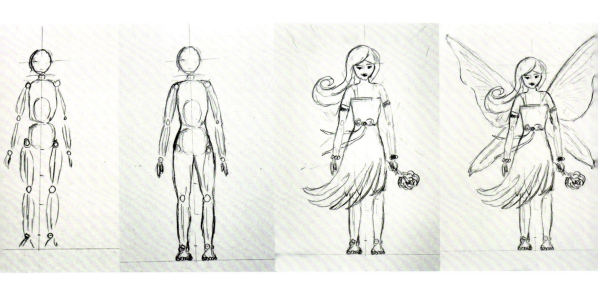

137

决定花仙子的种类 Deciding Type of The Fairy

因为生长环境不一样，不同种类的花仙子有着不同的外貌。

Different types of fairies have different appearances because they live in the different plant environments.

我决定我画的这个花仙子为槭树仙子。

I have decided that this type of fairy is a maple fairy.

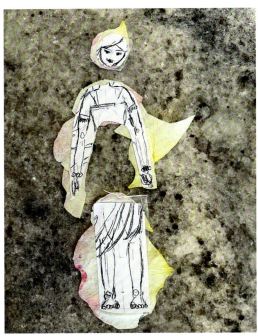 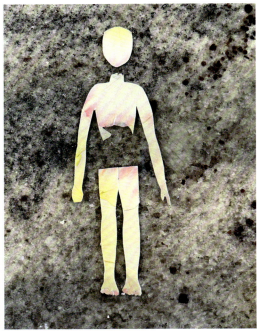

将图纸剪成段，然后在图纸背面贴双面贴，粘上玫瑰花瓣。根据图纸修剪所有部件。

Cut the drawing into sections and apply double sided adhesive to the back of the drawing. Glue rose petals. Trim all the parts according to the drawing.

幻想类 Fantasies

1

使用细酒精马克笔把眼睛、眉毛和嘴描好。

I used fine point alcohol markers to mark the eyes, eyebrows, and mouth.

2

剪香蕉皮为眼睛和眉毛。玫瑰花瓣为嘴唇。把它们贴好。

Cut banana skin to be eyes and eyebrows. Cut rose petal as lips. Glue them into position.

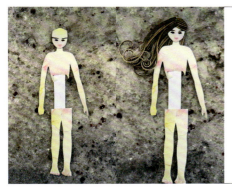

3

用纸条分别把头、身体和腿连接起来。使用铁树芽做她的长发。正反两面穿插使用铁树芽使得头发有层次。

Use paper strips to connect the head, legs, and the top of the body together.

I used young sago palm leaves as her hair. Alter the fount and back sides to make the high lights of the hair.

4

剪一片红色枫叶为她上衣和裙子衬里。用羽毛枫叶子为裙子的流苏。

Cut a red maple leaf to make her top and skirt lining. Use lacy maple as the fringe of her skirt.

5　为她戴上绿色的手镯和臂环。脚腕上戴上花环，发髻上插朵小花。剪一片枫叶为她的翅膀。

Put on green bracelets and an armband for her. Also put garlands on the ankle and a small flower on the side of hair face. Cut a yellow maple leaf to be her wings.

6　贴上土豆皮为槭树枝干。然后贴一些叶子。让仙子站在树枝上。最后把一朵马利筋花贴在她手里。

Glue potato skins to be the maple tree branches first. Then glue some maple leaves. Let the fairy stand on the tree branch. Finally, put a milkweed flower on her hand.

花仙子的动态 Drawing Fairy in Action

相比站直的花仙子，通常我更喜欢画仙子们的动态。这里有一个仙子坐着看往右边的模样。

Usually, I like to draw fairies in action rather than standing straight. Here is an example of step by step drawing of a fairy in sitting position looking out to the right.

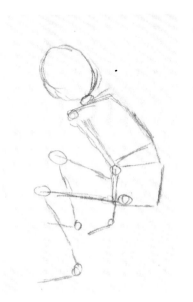

她比较瘦，而且她的翅膀也比较小。

She is slender and her wings are petite.

我设计了一个森林中她与蝴蝶互动的场景。

I designed a forest scene for her to play with butterflies.

仙子与童话故事有关。总有一个故事在制作幻想类画作时会激发我们的想象力。

Fairies are associated with a fairy tale. There is always a story. Making fantasy pictures invokes our imagination.

静 物 创 意 压 花 艺 术

丙烯泼画混合材料美人鱼
Acrylic Pouring Mixed Media Mermaid

幻想类 Fantasies

材料 Materials

1\. 纸张：混合媒介美术纸 140 lbs. 如果找不到混合媒介美术纸，可以选购表面比较光滑的水彩纸、Polycrylic或任何清漆。我们也需要双面贴。

Paper: Mixed media 140 lbs. If we cannot find mixed media paper, get smooth watercolor paper (140 lbs) and Polycrylic. We also need double sided adhesive.

2\. 丙烯颜色：任何牌子的都可以。我们需要白、天蓝和非常浅的蓝。如果想要金属色彩，则必须选购专业级的丙烯。手工级的丙烯金属色素不够浓。如果喜欢，还可以加入海洋绿。

Acrylic colors: any brand color would work. We need white, pale blue, and very light blue. If we want to have metallic color, you will have to invest on a good professional paint. The craft ones don't have enough pigment. If you want to add more colors, ocean green is a good choice.

3\. 市面上有已经混合好的流体画颜料出售。可以购买那些，跳过混合步骤。
白乳胶和硅胶油。可以上网选购或到体育用品店购买。这是用来为跑步机的带子润滑的。也有专门为流体画准备的硅胶油。

There are pre-mixed ready to pour colors available. We can get those and skip all the mixing steps. There are even coastal wave ready to pour sets available.

White glue, and silicone oil. We can find silicone oil either online or sports equipment store. It is used for treadmill belt lubrication.

There are also silicone oil for acrylic pouring available now.

145

普通工具如镊子、剪刀和热风枪。

Usual tools such as tweezers, scissors, and heat gun.

压花花材：浅粉或奶油色大玫瑰花瓣；小康乃馨、中小双色玫瑰或粉红边绣球；铁树芽或深色卷藤；芍药、玫瑰、圣诞红、其他大型好看的花瓣或叶片；原色半叶脉；木兰花瓣或其他咖啡色叶子。

Pressed materials:

Cream or light pink larger rose petals, small carnation or small/medium two tone rose petals or pink edge hydrangea; sago palm or other dark color tendrils; peony or rose or poinsettia or other pretty color larger petals or leaves; skeleton leaves (original color), magnolia or other brown petal/leaves.

准备好工作台面 Prepare Work Surface

丙烯泼画有点脏，所以需要找一个大纸盒，或塑胶布把桌面保护好。

Acrylic pouring is somewhat messy, so we need to line your workspace with a cardboard, or a piece of freezer paper, or plastic sheets.

幻想类 Fantasies

纸张 Paper

如果使用的是混合媒介美术纸，跳过下面这步。

使用纸胶带把背景纸固定在硬板上。在纸面上涂一薄层polycrylic或其他清漆以密封。等10分钟后再涂一层。然后等待至少30分钟让纸干透。最好就是把干透的纸夹在大的厚书里面过夜，这样纸张会非常平整。

If we are using the mixed media paper, then skip the following step.

Use masking tape or inexpensive washi tape to secure the background paper onto a piece of hardboard. Apply one thin coat of polycrylic onto the watercolor paper to seal it. Wait for 10 minutes and apply another coat. Let it dry completely for at least 30 minutes. It is a good idea to place the background paper into a large thick book overnight so the paper will be very flat.

垫高 Padding

我们需要把背景纸张垫高，这样多余的丙烯色彩会流下去，而不会回流污染泼画。先用纸胶带把背景纸贴在硬板上。然后用纸胶带把我在一元商店买的一个小的厨房用架子在硬板后面固定好。

We will need to raise the background paper high so excessive acrylic will flow off the paper and not contaminate the pour. First, tape the background paper onto a piece of cardboard.

 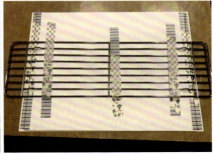

I used a small cooking rack from the dollar store. Using masking tape or washi tape to secure the rack onto the back of the hardboard.

水平 Level

利用手机的水平app，你可以轻松地知道桌面是否水平。如果需要，可以折叠纸张把其中的桌脚垫一下。桌面必须水平才能令泼的丙烯在干燥期间保持你设计的模样。

You will need to make sure the working surface is leveled. My cellphone has a levelling app. I can use it to make sure the surface is leveled. Adjust your table legs by padding a few sheets of paper if necessary. The Surface must be level so the pouring will stay as you desire during drying period.

准备好泼的丙烯 Prepare Colors for Pour

如果采购了混合好的流体画颜料，跳过这步。将一份白胶、一份丙烯和一份水混合。用冰棍棒搅拌均匀。这是一个通用公式。必须根据颜色调整水和胶水，因为胶水和丙烯酸的稠度因品牌而异。稠度应该是提起冰棍棒，棒上保留一些色彩合剂，多余的合剂形成一条直线流下，而且流畅。如果流速过慢，则加少量水。如果流速过急，冰棍棒上不能保留什么色彩合剂，则加入胶来调节。

除白色以外，加入两滴硅油至所有混合颜色。请确保不要添加太多的硅油，需要的就是两滴。不要用力过度挤压瓶子。硅油对于泼画中形成细胞是必需的。

If you got pre-mixed ready to pour colors, skip this step.Mix one part of white glue, one part of acrylic, and one part of water. Stir well with a popsicle stick. This is a general formula. We have to adjust the ratio of water and glue for our color because thickness of the glue and acrylic is different from brand to brand. The thickness should be that some color remains on the stick but extra color forms a line of flow when pulling the popsicle stick up. If the flow is forming a line but drops slowly, then add a little water at a time until the flow is smooth. If it is too watery that no color is sticking on the stick, add more glue.

Add two drops of silicone oil to all mixed colors except white. Please make sure not to add too much silicone oil. All we need is two drops. Don't squeeze the bottle too hard. Silicone oil is necessary for creating cells in the pouring.

拿一个空杯子，倒入一些白色，然后各种颜色。重复更多的颜色。将足够的颜色倒入杯子中。确保可以覆盖整个表面。把冰棒棍插入，慢慢地转两个圈（不能多搅拌）。

Take an empty cup, pour some white, then each color, and then add more colors. Pour enough colors into the cup. Make sure we will not be running out of color to cover the entire surface. Insert the popsicle stick and spin slowly for two complete circles (Do not spin more than twice).

泼画 Pouring

将背景与厨房架子放在工作台上。戴一次性手套。在边缘倒一圈白色。这是让颜色流动时能够流动得顺畅。然后将丙烯混合物全部倒在背景表面。可以使用冰棒棍帮助引导丙烯流动，但流动应该主要通过倾斜背景表面来完成。旋转背景，使丙烯覆盖表面，而不是全部流下去。可以将流下去的丙烯再舀回到背景表面，但不要让这种情况发生超过两次。否则背景会看起来脏，而不是清晰干净的图像。在丙烯覆盖整个表面之后，使用热风枪加热表面，以使细胞出现。不能过长时间使用热风枪，否则会导致丙烯过热并产生气泡。在一个区域做短暂的加热以便看到细胞形成。让背景过夜干透。

Place the background with the rack on the work surface. Wear disposable gloves. Pour some white colors on all sides. This is to guide the color flow onto the edge. Now, pour the entire content of the paint mix onto the surface. Use the popsicle stick to help guide the paint to flow but the flow should be mostly done by tilting the surface of the background. Rotate the background so we have paint covering the surface and not flowing off. It is fine to scoop excess paint back onto the surface of the background but do reapply more than two times. Otherwise, we will have a muddy background instead of clear and distinct paints. Use a heat gun to apply heat on the surface to induce the silicone oil to pop up after you have the entire surface covered. We do not want to blow the heat gun for too long or you will cause the acrylic to overheat and create bubbles. Do very short bursts of heat over the area just enough to see cells forming. Let the background dry overnight.

制作压花美人鱼 Making Pressed Flower Mermaid

美人鱼是一个幻想的生物，本篇样品建议了一些材料，但我们可以根据喜好使用任何材料。我创作人物的方式是通过观察人物模型，然后勾画线条以形成人物。我喜欢有合适比例，看起来真实的画。欢迎大家使用我的图样。如果您使用来自网络资源或图书的图片，请确定版权可以允许你这样做。

Mermaids are a fantasy creature. My samples suggest some materials you can use but we can use any materials as you see fit. The way I create my figures is by looking at my figure model and then sketch the lines to form the figure. I like to have figures that are real looking with the right proportions. You are welcome to use my drawings. If you are using an image from online sources or from a book, make sure you are permitted to do so.

制作美人鱼 1 Procedure for Making Mermaid One

1 把反转的美人鱼图打印出来，并且把身体大致剪下来。放在双面贴上。

Print a flipped mermaid image. Cut the body out (not too closely) and place it on double sided adhesive.

2 剪掉多余的双面贴，把离型纸撕下来，然后粘在一片大玫瑰花瓣上。

Cut off extra adhesive. Take the protective paper off and glue it onto a large rose petal.

3 根据图纸剪下身体形状及胳膊的线条。粘贴剪得非常细的粉红色花瓣或细长的卷须来定义手臂。

Cut the body shape out according to the drawing. Cut the lines for arms. Glue very thin lines of pink petal or thin lines of tendrils to define her arms.

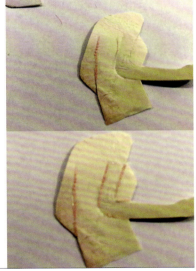

4 把腰部以下大致剪下来，放在双面贴上。剪掉多余的双面贴。

Cut the tail off roughly from waist down and place it on the double sided adhesive. Cut the extra adhesive off.

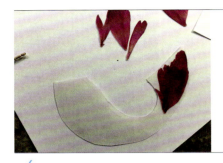

把双面贴的离型纸撕掉。选择一些芍药花瓣，贴在尾巴上。

Take the protection paper off from the double sided adhesive. Select some peony petals for the tail and glue on.

只使用迷你康乃馨花瓣的上边缘，在尾部粘上。选择较小的花瓣，堆叠和穿插使其像鱼鳞一样。

Use just the upper edge of mini carnation petals, glue on the tail. Select smaller petals. Stack and interweave, making it appear like fish scales.

把尾部翻过来，剪掉多余的迷你康乃馨部分。

Flip the tail over and trim the carnation portion of the tail.

翻回正面，根据芍药花瓣的形状将尾巴修剪好。如果材料不如芍药那么像鱼尾，也可以根据图纸进行修剪。

Flip the tail back and trim the peony portion of the tail according to flower shape. (If our material is not neat like peony, we can trim according to the drawing).

幻想类 Fantasies

把头部贴在双面贴上,将脸和少许头发剪下。

Glue the head onto double sided adhesive. Cut the face with a little of the hair out.

贴一小片玫瑰花瓣在脸上,剪掉多余的花瓣。

Glue a small piece of rose petal on the face. Trim to shape.

用小段纸条把头、身体和尾部连接起来。

Assemble the head, body and tail together and use a small strip of paper to secure all the parts together.

把头部的双面贴离型纸撕下,排列铁树芽为头发。

Take off the protection paper, line sago palm leaves for hair.

13

把头型修剪一下，然后把铁树芽的卷粘在身体前面。

Trim the head shape. Glue some sago palm tendrils for long hair in front of the body.

14

将一小片双面胶粘贴到一片银叶背面上。用一打洞器打一个半英寸（11mm）的圆圈。也可以绘制和剪这样的圆圈。

用镊子撕下小块自然色叶脉（或两种颜色的棕色叶）。在圆圈的右下方添加阴影，这是吸引美人鱼注意的巨型珍珠。

Glue a small piece of double sided adhesive onto a piece of silver leaf. Use a punch to punch a half inch circle. We can draw and cut such circle too.

Use tweezers to tear small pieces of natural color skeleton leaf (or two colors of thin brown leaves). Add shade on the lower right side of the circle. This is a giant pearl that attracts the mermaid's attention.

现在我们可以把美人鱼放在背景上。我使用15.5cm×23.5cm。你也可以使用12.7cm×17.8cm。

Now it is time to place the mermaid onto the background paper. I use 6"×8". We can use 5"× 7" also.

我们可以把她的头发、手臂、手腕和腰加上花饰。我使用的是香雪球的小花。我们还可以在她尾巴后面加上鳍。

Add flower bands for her hair, arm, wrist, and waist. Small alyssum flowers works well. We can also add fins on the back of her tail.

制作美人鱼 2
Procedure for Making Mermaid Two

制作潜水美人鱼的步骤与小美人鱼相同。她的面部特征非常细微。我用了小片深色卷须，我发现佛手瓜卷须非常有用，自然发黑。让我们再次快速回顾一下这些步骤。

Making the diving mermaid is the same as the smaller mermaid. Her facial feature is very delicate. I have used tiny pieces of dark tendrils. I find Chayote melon tendrils very useful. It presses dark naturally. Let's do a quick review of the steps again.

幻想类 Fantasies

1 打印后面反转图像。使用双面胶粘贴玫瑰花瓣背面的部分，剪出形状。

Print the reverse image of the mermaid. Use double sided adhesive to glue the parts on the back of the rose petal. Cut the shape out.

剪下半身的形状（不要太靠近边缘）。使用合适的花瓣作为鱼鳞。

Cut the general shape (not too close to the edge) of the lower body. Use suitable petals for the scales.

3 使用相同的双面胶的技术来切割尾部形状。在尾巴上粘上一些小花瓣，给它一些纹理。组装下半身、尾巴和鳍。

Use the same technique using double sided adhesive to cut the tail shape. Glue some small petals onto the tail to give it some texture. Assemble the lower body, the tail and fin.

根据图纸将所有零件组装在一起。
使用深色卷须作为眼睛和眉毛，小小的芍药作为嘴唇，小小的玫瑰花瓣作为鼻子。

4

Assemble all the parts together according to the drawing.
Use dark tendrils for eyes and eyebrows, a tiny piece of peony for her lips, and a tiny piece of rose petal for her nose.

5 将美人鱼粘到背景上。为头发的细节添加深色卷须。为她的手臂、手腕和腰部上的首饰添加米香花。

Glue the mermaid onto background. Add dark tendrils for detail of the hair. Add rice flowers for her jewelries on her arm, wrist, and waist.

美人鱼另一例
Another Example of Mermaids

头发用的是变色的黄色菊花花瓣。

The hair is made of faded yellow chrysanthemum petals.

后记

Postface

　　历时三年，重写数次之后，我终于把这套书写完了。
　　非常感谢中国林业出版社的编辑印芳一直以来的支持。
　　非常感谢我的家人给与我的鼓励和支持，感谢我儿子帮忙为我校正英文版本，感谢我先生给我写作方向的一些启发。
　　感谢国际压花协会的老师们和会员们给我的鼓励，并且在日常的交流中、各种活动中还有不同的课程中给我很多启发。
　　我希望这套书可以为每一位喜欢压花艺术的读者提供一些思路，从而创作出具有个人风格的艺术。也为大家介绍各种压花艺术独特的技巧和方法。

I finally finished writing this set of books after revising it for three years.

Thank you very much Yin Fang, editor of China Forestry Publishing House, for your continuous support.

I thank my family very much for the encouragement and support. Thanks to my son for helping me correct the English version. Thanks to my husband for giving me some inspiration on writing direction.

Thanks to the teachers and members of the World Wide Pressed Flower Guild for their encouragements and inspirations in daily communication, various activities, and in different classes. I hope my book will provide some ideas for every artist and to create art with their own personal style. I hope to also introduce some unique techniques and methods of pressed flower art.

2021.04